SpringerBriefs in Energy

More information about this series at http://www.springer.com/series/8903

Mrinmoy Majumder · Apu K. Saha

Feasibility Model of Solar Energy Plants by ANN and MCDM Techniques

 Springer

Mrinmoy Majumder
Department of Civil Engineering
National Institute of Technology Agartala
Agartala
India

Apu K. Saha
Department of Mathematics
National Institute of Technology Agartala
Agartala
India

ISSN 2191-5520　　　　　　　　ISSN 2191-5539　(electronic)
SpringerBriefs in Energy
ISBN 978-981-287-307-1　　　　ISBN 978-981-287-308-8　(eBook)
DOI 10.1007/978-981-287-308-8

Library of Congress Control Number: 2016936958

Printed on acid-free paper

This Springer imprint is published by Springer Nature
The registered company is Springer Science+Business Media Singapore Pte Ltd.

Preface

The large-scale urbanization and technological advancements have induced high demand for energy in nearly every part of the World. As a result the finite source of fossil fuels which is the main source of supply has failed to satisfy this growing need of energy. Not only electricity, fuel is required to run automobiles, maintain industrial output, and for many other purposes.

The necessity of more fuels imbibed the need of an alternative source of energy which is available infinitely but not expensive to utilize as electricity or fuel. The solar energy among many other renewable sources of energy is one of the reliable most option in this aspect.

The solar power utilization is costly. The conversion efficiency lies within 40 %. The resource availability is during the daylight hours only. All these constraints except conversion efficiency changes with change in the location. The cost of installation, maintenance and labour charge also varies with location.

That is why if optimal location for solar energy utilization can be identified then quality, quantity and time of production will be ensured to be maximum among all the available alternatives.

The present investigation is an attempt to provide a methodology for site selection which will be both objective and cognitive. Multi-Criteria Decision-Making and new variant of Neural Networks like Group Method of Data Handling was utilized to imbibe objectivity and cognitivity into the procedure respectively.

Chapter 1 introduces the justification of the present investigation and in Chap. 2 the importance of solar energy was highlighted. Chapter 3 states the strength, weakness and application of Multi-Criteria Decision-Making in decision-making problems.

Chapter 4 depicts the advantage, disadvantage and applicability of new Artificial Neural Networks like Group Method of Data Handling and the detail methodology and its way of implementations in the present study was also described in Chap. 5.

The sixth chapter describes the results derived from the Analytical Hierarchy Process Multi-Criteria Decision-Making process and the predictive models developed with Group Method of Data Handling. The result of the application of the developed indicator in 12 different cities for knowing their feasibility as a location of solar power generation was also discussed in this chapter.

The seventh chapter concludes the study with highlighting the strength, weakness and future scopes of the study.

Acknowledgments

We would like to tender our gratitude to all those people who saw us through this book; to the persons who provided support, discussed, read, wrote, offered suggestions, allowed us to quote their remarks and assisted in the editing, proofreading and design.

We would like to thank the reviewers, publishers and production assistants for enabling us to finish the manuscript and publish the same in time. Above all we want to thank our family, colleagues and friends who supported and encouraged in the long and difficult journey of the book preparation. Their contribution and sacrifice will never be forgotten.

Last but not least, We sincerely convey my apologies to all those who have been with us during the development of this manuscript but whose names we have omitted inadvertently.

Agartala   Mrinmoy Majumder
February 2016   Apu K. Saha

Contents

Contents

Chapter 1
Introduction

Abstract The scarcity in the supply of energy and rapid urbanization all over the world has enforced to look for alternative sources of energy. Solar energy is one of the most popular and reliable source of energy which is infinitely available and can become a possible substitute to fossil fuels. As most of the factors on which the usability of solar energy depends varies with location, selection of suitable sites for solar power plant is the key for optimal utilization of the potential. Present study uses AHP and GMDH to identify feasible sites for installation of solar power plants.

Keywords Solar · Site selection · AHP

The large scale urbanization and advances in technology has increased the demand for energy manifold. It was seen that in Kenya an increase in 3.10 % of the population will increment the power consumption by 1.2 % per year (Christina 2012). Globally the ratio of population, energy use and GDP was found to be 1.18, 2.33 and 3.51 % respectively in the year of 2010 (Tverberg 2012).

The population is under rapid growth and so is the energy consumption. But on the other hand countries with higher energy consumption is also the higher emitter of Green House Gas (GHG), one of the major factor of climate change. As for example, Australia, Canada and United States (World Bank 2011) are among the countries having highest energy consumption (as per kilogram of oil equivalent) but these countries are also listed as 4th, 6th and 7th ranked countries with respect to per capita emission of GHG gases globally (Baumert et al. 2005).

The global ration of demand for energy is around 12,000 Million Tonnes oil equivalent (IEA 2010) and it is expected to grow by 37 % up to the year of 2040 (IEA 2014). The 85 % of this demand is supplied from fossil fuel and nearly 15 % is the contribution of renewable energy sources. In the year of 2013 the contribution of renewable and nuclear energy source was 24.5, 12.1 and 2.2 % respectively in Canada, United States and Australia. In 2012, contribution of alternative sources were 14.5, 4.3 and 2.9 % in Brazil, China and India respectively whereas globally the contribution was 8.5 % (IEA Statistics 2014).The contribution from renewable is expected to increase to nearly 18 billion tonnes of oil equivalent in the year of 2035

© The Author(s) 2016
M. Majumder and A.K. Saha, *Feasibility Model of Solar Energy
Plants by ANN and MCDM Techniques*, SpringerBriefs in Energy,
DOI 10.1007/978-981-287-308-8_1

but share of contribution may increase 0.45 billion tonnes of oil equivalent compared to the share of coal,oil and gas which will be equal to around 7–8 for coal and oil and 6–7 billion tonnes of oil equivalent for gas (IEA 2007).

Globally the demand for electricity is increasing by 2.5 % per year whereas contribution from renewable energy is incrementing by 20 %. Among the many sources of renewable energy, contribution from Solar energy to satisfy global demand was found to be 16.2 % for photo-voltaic and 9.2 % for thermal electricity. (Bastos n.d).

But the problems with solar energy lie with the cost and efficiency of procedures to extract electricity from the photons available in the solar radiation.

Although, the maximum efficiency found for conversion is around 40 % the main obstacles to wide scale utilization of solar energy remains its dependency on location.

1.1 Solar Energy

The basic working principle of the conversion of solar power to usable forms of energy is given in Fig. 1.1.

The factor which increases or decreases the suitability of a location for installation of soler power plant is depicted in Fig. 1.2.

1.2 Objective of the Present Study

The objective of the present investigation is to develop an indicator which will represent the feasibility of solar energy plants in locations of installation.

In this regard it was also aimed that the priority and priority values of different parameters with respect to solar energy utilization can also be identified.

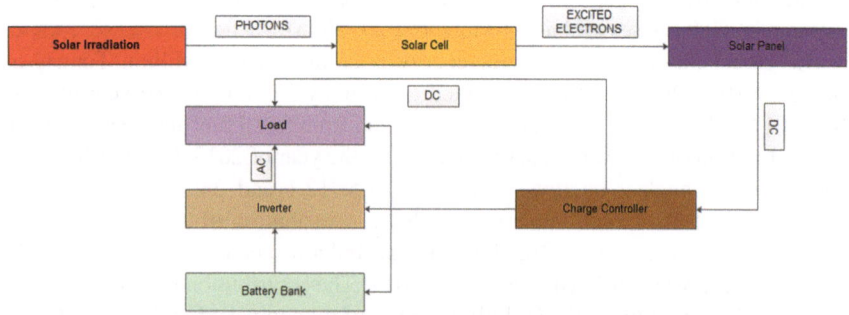

Fig. 1.1 Figure showing the working principle of solar energy convesrsion

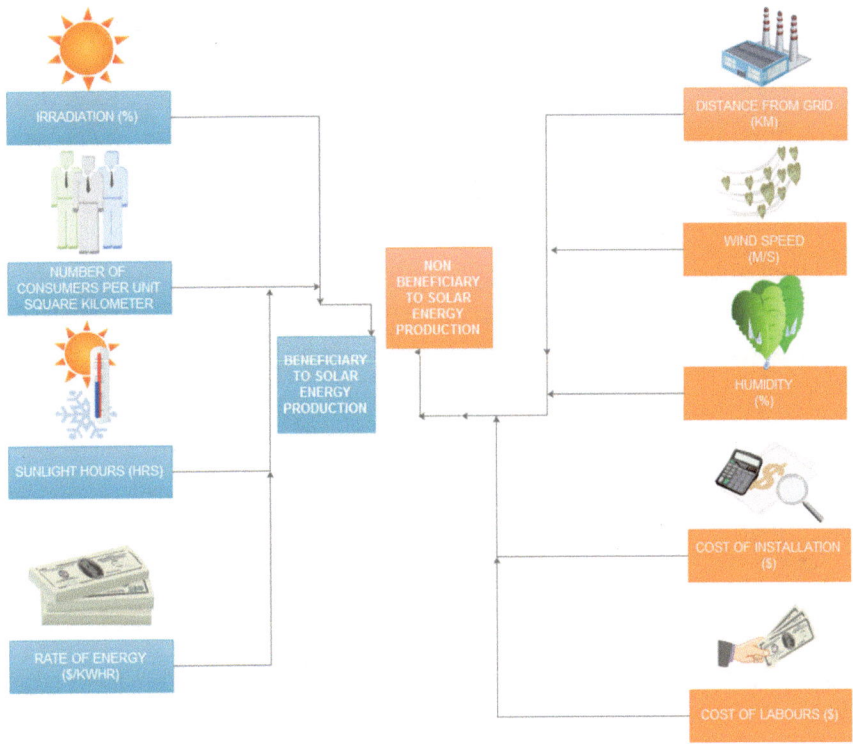

Fig. 1.2 Factors affecting the electability of a location for installation of solar energy power plant

The said indicator can become an easy tool for identification of suitable lands to install solar energy plant.

1.3 Case Studies

In total the indicator was applied to find the location suitability of the available space for 12 different cities in the World. Figure 1.3 depicts the locations of the cities where the indicator was applied to find the suitability of installation of solar power plant.

Two cities were from Indian subcontinent, two are from Oceania, Latin and Central America. The cities were also selected from Africa and European countries.

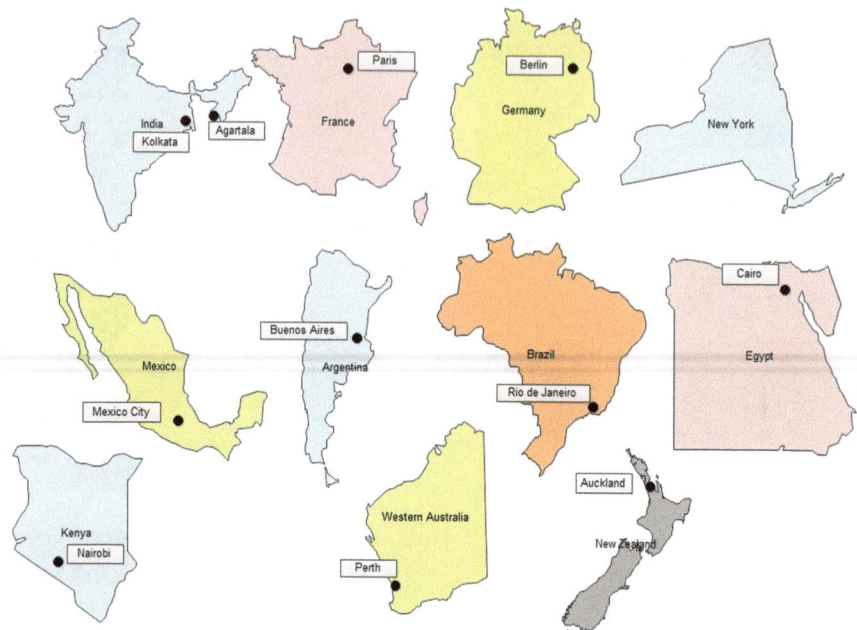

Fig. 1.3 Figure showing the location of the cities which are selected as study area for the present investigation

References

Bastos CP, (unknown), Contribution of Solar and Wind energy to the World Electrical energy demand, Retrieved from www.sefidvash.net/fbnr/pdfs/Solar_and_Wind_Energy.pdf on 8 Feb 2016

Baumert KA, Herzog T, Pershing J (2005) Navigating the numbers: greenhouse gas data and international climate policy. World Resources Inst

Christina (2012) Energy in Kenya and the potential for renewable, Retrieved from http://investeddevelopment.com/blog/2012/08/energy-in-kenya-and-the-potential-for-renewables/ on 6 Feb 2016

IEA (2014) World Energy Outlook 2014, Executive summary, Retrieved from www.iea.org/textbase/npsum/weo2014sum.pdf on 8 Feb 2016

IEA Statistics (2014) Alternative and nuclear energy (% of total energy use), Retrieved from http://www.iea.org/stats/index.asp on 8 Feb 2016

IEA 2010 World Energy Outlook 2010, Executive summary, Retrieved from www.iea.org/textbase/npsum/weo2010sum.pdf on 8 Feb 2016

IEA 2007 World Energy Outlook 2007, Executive summary, Retrieved from www.iea.org/textbase/npsum/weo2007sum.pdf on 8 Feb 2016

Tverberg G (2012 The long-term tie between energy supply, population, and the economy, Our Finite World, Retrieved on 6 Feb 2016

World Bank (2011) "Energy Use Per Capita". World Development Indicators, Retrieved from http://data.worldbank.org/indicator/EG.USE.PCAP.KG.OE on 8 Feb 2016

Chapter 2
Solar Energy

Abstract Solar energy mainly depends on insolation intensity and duration of sunlight hours. Both of this varies with location. That is why if suitable location can be identified, optimal utilization of the resources is possible. Solar energy is a popular alternative and renewable form of energy source which can easily be converted to utilizable forms of energy. Although conversion efficiency and cost is not conducive enough to designate this form of energy to substitute the conventional energy sources. But depending on location cost can be reduced and amount of energy can be increased.

Keywords Renewable energy · Conversion cost · Benefits

The solar energy is deemed to be one of the most favourable alternative forms of energy which can replace the conventional energy sources.

The strength, weakness, opportunity and threat or SWAT analysis of solar energy is detailed in Fig. 2.1.

It seems that the major problem of electricity production from solar energy is the availability of sunlight. The amount of electricity that can be produced depends on duration and amount of available sun light. Both of which are location dependent.

Again, the availability of un-shaded area is also an important factor. More the area more panels can be installed to convert solar energy to electricity. The non-shaded area also varies with location.

The cost of installation and maintenance also changes with locations. Another important factor which changes with location is distance from grid or consumers. The wind speed and humidity increases the maintenance requirement of solar panels which are also varies with locations.

That is why it can be seen that most of the factors which influences the efficiency of solar energy production changes with location. The selection of location thus becomes extremely important in the success of a solar power plant.

© The Author(s) 2016   5
M. Majumder and A.K. Saha, *Feasibility Model of Solar Energy*
Plants by ANN and MCDM Techniques, SpringerBriefs in Energy,
DOI 10.1007/978-981-287-308-8_2

Fig. 2.1 Figure showing the
SWAT analysis of solar
energy

2.1   Problems

The previous studies of selection of suitable location for installation of solar power
plant are given in Table 2.1.

Although the studies used various tools and techniques to identify optimal
location for solar power production, most of the studies have a common disadvantage. It does not selected the factors which influence the selection by using an
objective method. Also the selection of factors is based on expert suggestion or
views from the stakeholders or experience of the engineers. In most of the cases
involving manual interpretations introduce bias and decision changes with decision
makers.

The application of both Multi Criteria Decision Making (MCDM) and
Artificial Neural Network (ANN) is omnipresent in this type of studies.
Although MCDM is used to select location for solar power plant but Group
Method of Data Handling, a new variant of ANN, has never been applied in this
aspect.

Table 2.1 Table showing the previous studies which identified location for solar power plant with the help of different tools and techniques

Location	Factors considered	Techniques utilized	Type of plant	Authors
India	Radiation, levelized unit cost of electricity	System advisor model	Parabolic trough concentrator (ptc) and linear fresnel reflector (lfr) based solar thermal power plants	Sharma et al. (2016)
Brazil	Location, area and shape	RET screen or homer	Small scale solar and wind power	Ribeiro et al. (2016)
India	Emission reduction constraint	Differential evolution	Solar PV	Rajesh et al. (2016)
Italy	Standard deviation of energy balance and storage requirement		Solar PV and ROR HPP	Francois et al. (2016)
Turkey	Quality of terrain, local weathering factors, proximity to high transmission capacity lines, agricultural facilities and environmental conservation issues	AHP-GIS	Solar farms	Uyan (2013)
India	Techno-economical and environmental factors	RET screen on the basis of viability indicators like internal rate of return (IRR), net present value (NPV), cost of electricity (CoE), benefit–cost (B–C) ratio	Grid connected PV	Jain et al. (2011)
Spain	Social, economic, technological, and environmental factors	AHP-VIKOR	Solar Thermo Electric	San Cristóbal (2011)
Spain	Project execution delay and stoppage risk	Analytical network process	Solar PV	Aragonés-Beltrán et al. (2010)
European Union	Environment, orography, location, and climate factors	Analytical hierarchy process with geographical information system	Grid connected PV	Carrion et al. (2008)
Northern Africa	Direct solar radiation, geographical conditions (land slope, land cover, distance from cooling water resources, etc.), infrastructure (pipelines, electricity grids, streets etc.) and the configuration and performance of a selected solar thermal power plant concept	STEP model, satellite data and GIS	Solar thermal power stations	Broesamle et al. (2001)

References

Aragonés-Beltrán P, Chaparro-González F, Pastor-Ferrando JP, Rodríguez-Pozo F (2010) An ANP-based approach for the selection of photovoltaic solar power plant investment projects. Renew Sustain Energy Rev 14(1):249–264

Broesamle Hartmut, Mannstein Hermann, Schillings Christoph, Trieb Franz (2001) Assessment of solar electricity potentials in North Africa based on satellite data and a geographic information system. Sol Energy 70(1):1–12

Carrión JA, Estrella AE, Dols FA, Toro MZ, Rodríguez M, Ridao AR (2008) Environmental decision-support systems for evaluating the carrying capacity of land areas: Optimal site selection for grid-connected photovoltaic power plants. Renew Sustain Energy Rev 12(9):2358–2380

Francois Baptiste, Borga Marco, Creutin Jean-Dominique, Hingray Benoit, Raynaud Damien, Sauterleute Julian-Friedrich (2016) Complementarity between solar and hydro power: sensitivity study to climate characteristics in Northern-Italy. Renew Energy 86:543–553

Jain Amit, Mehta Rajeev, Mittal Susheel K (2011) Modeling impact of solar radiation on site selection for solar PV power plants in India. Int J Green Energy 8(4):486–498

Ribeiro AED, Arouca MC, Coelho DM (2016) Electric energy generation from small-scale solar and wind power in Brazil: the influence of location, area and shape. Renew Energy 85:554–563

Rajesh K, Bhuvanesh A, Kannan S, Thangaraj C (2016) Least cost generation expansion planning with solar power plant using differential evolution algorithm. Renew Energy 85:677–686

Sharma Chandan, Sharma Ashish K, Mullick Subhash C, Kandpal Tara C (2016) A study of the effect of design parameters on the performance of linear solar concentrator based thermal power plants in India. Renew Energy 87:666–675

San Cristóbal JR (2011) Multi-criteria decision-making in the selection of a renewable energy project in Spain: the Vikor method." Renew Energy 36(2):498–502

Uyan Mevlut (2013) GIS-based solar farms site selection using analytic hierarchy process (AHP) in Karapinar region, Konya/Turkey. Renew Sustain Energy Rev 28:11–17

Chapter 3
Multi Criteria Decision Making

Abstract Multi Criteria Decision Making (MCDM) is one of the technique which is used to select most optimal alternative with respect to multiple criteria for a specific goal. The method provides objectivity and compares the alternative relatively to estimate the priority value of the alternatives. Based on the priority value the optimal alternative is identified and selected as the option which can achieves the decision objective.

Keywords Multi criteria decision making · Priority value · Objective decision making

The Multi Criteria Decision Making (MCDM) method is popular for objectively solving decision making problems.

The Fig. 3.1 shows the working principle of MCDM methods and discuss about the different types of MCDM that are widely used.

The strength and weakness of the MCDMs are highlighted is also depicted in the same figure.

The MCDM techniques are nowadays widely applied and used to solve various decision making, optimization and predictive problems. Table shows some of the applications in the energy field.

Many different kinds of MCDM techniques are used nowadays for solving various decision making problems. Among them Analytical Hierarchy Process (AHP), Fuzzy Logic Decision Making (FLDM), Weighted Sum Method (WSM), Weighted Product Method (WPM), EVAMIX and ELECTRE is one of the few which are widely popular for their reliability.

© The Author(s) 2016
M. Majumder and A.K. Saha, *Feasibility Model of Solar Energy
Plants by ANN and MCDM Techniques*, SpringerBriefs in Energy,
DOI 10.1007/978-981-287-308-8_3

	Methodology	Characteristic	Example Methods	Strength	Weakness	Applications
1 Define the Goal of the decision	Selection and determination of decision goal by discussion with experts,stakeholders and even by voting	Specific, Achievable and Realistic	.		.	
2 Selection of Criteria	Based on the decision goal,criteria for comparison of alternatives are selected based on expert,stakeholder,literature etc. surveys	Specific, Realistic and Related	.		.	
3 Selection of Alternatives	Alternatives will be either available or selected	Mensurable, Coherent to goal, real and comparable	.		.	
4 Comparison and rating of alternatives as per its importance to the decision goal	The importance of alternatives are identified with the help of literature,expert,stakeholder/der etc. surveys and by voting or scoring methods	Rating is performed either by a pre-specified scale or by the magnitude	Delphi SMART etc.	Responses remains confidental Minimum time required to complete survey Cost effective and flexible/adaptable,	Time delays for data-collection Respondents may become non-committed and unreliable or biased. Requires skill in writing.	Linstone and Turoff, 1975 Cables et.al., 2016
5 Application of aggregation methods to find the equivalent weights of each alternatives with respect to each criteria	Aggregation methods are of different types.Some of them averages/geometric mean/products to find the equivalent weight	The output is generally normalized before being used for ranking the alternatives.	.		.	
6 Ranking of alternatives based on the equivalent weights	The results from the aggregation method is used to rank the alternatives in a descending manner	Can be ranked according to equivalent weights and top ranked alternative is selected or one alternative can outrank the other by comparing their equivalent weights	TOPSIS AHP ANP ELECTRA PROMETHEE	Ease of Use Flexibility Quantification Sometimes comparison yields relative importance of the alternatives	Produce unrealistic results if number of alternatives is small or rating of alternative is incorrectly performed. Decision may change with change in methods.	Xu et.al.,2016 Akash et.al.,1999 Strantzali and Aravossis,2016 Ishizaka et.al.,2016

Fig. 3.1 Figure depicts the working principle, types, strength, weakness and applications of multi criteria decision making

3.1 Analytical Hierarchy Process

The AHP method is an example of compensatory MCDM method which compares each alternative with the other alternatives with respect to each of the criteria and then finds the equivalent weight of importance for each of the alternatives which are derived by aggregating the individual importance of the alternatives found while comparing based on each of the criteria. Figure 3.2 highlights the applications of AHP.

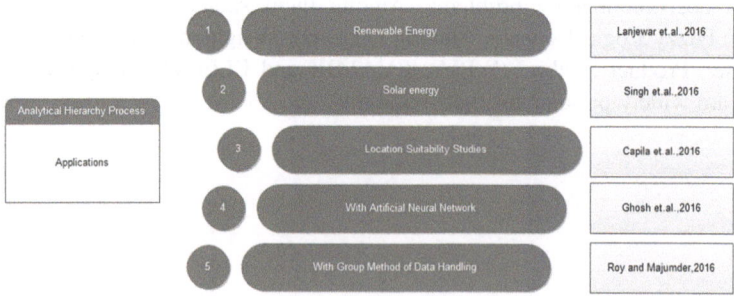

1	Renewable Energy	Lanjewar et.al.,2016
2	Solar energy	Singh et.al.,2016
3	Location Suitability Studies	Capila et.al.,2016
4	With Artificial Neural Network	Ghosh et.al.,2016
5	With Group Method of Data Handling	Roy and Majumder,2016

Analytical Hierarchy Process — Applications

Fig. 3.2 Figure showing the applications of AHP in decision making tasks from different related fields

The strength of this method included but not limited to the fact that it is flexible, easy to use, gives relative importance and accepts both qualitative and quantitative alternatives. But the disadvantage of this method is that it does not clearly define the rating of similarly important variable. Also the hierarchical structure of the method prevents it to make a decision bidirectional.

The present study utilized AHP for its ability to consider both quantitative and qualitative variables and the time it takes to make a decision.

The study had compared the results from Analytical Network Process (ANP) (Saaty 1978) with the AHP results for selection of the optimal results for the study. ANP is a method which is bidirectional, i.e., it considers the importance of alternative with other alternatives based on criteria and vise-versa.

References

Akash Bilal A, Mamlook Rustom, Mohsen Mousa S (1999) Multi-criteria selection of electric power plants using analytical hierarchy process. Electr Power Syst Res 52(1):29–35

Cables E, Lamata MT, Verdegay JL (2016) RIM-reference ideal method in multicriteria decision making. Inf Sci 337:1–10

Capilla JJ, Carrión JA, Alameda-Hernandez E (2016) Optimal site selection for upper reservoirs in pump-back systems, using geographical information systems and multicriteria analysis. Renew Energy 86:429–440

Dong L, Liang H, Gao Z, Luo X, Ren J (2016) Spatial distribution of China? s renewable energy industry: regional features and implications for a harmonious development future." Renew Sustain Energy Rev

Gherboudj Imen, Ghedira Hosni (2016) Assessment of solar energy potential over the United Arab Emirates using remote sensing and weather forecast data. Renew Sustain Energy Rev 55:1210–1224

Ghosh Soumya, Chakraborty Tilottama, Saha Satyabrata, Majumder Mrinmoy, Pal Manish (2016) Development of the location suitability index for wave energy production by ANN and MCDM techniques. Renew Sustain Energy Rev 59:1017–1028

Ishizaka Alessio, Siraj Sajid, Nemery Philippe (2016) Which energy mix for the UK (United Kingdom)? An evolutive descriptive mapping with the integrated GAIA (graphical analysis for interactive aid)–AHP (analytic hierarchy process) visualization tool. Energy 95:602–611

Lanjewar P, Rao R, Kale A, Taler J, Oclon P (2016) Evaluation and selection of energy technologies using an integrated graph theory and analytic hierarchy process methods. Decision Sci Lett 5(2):237–348

Linstone Harold A, Turoff Murray (eds) (1975) The Delphi method: Techniques and Applications, vol 29. Addison-Wesley, Reading, MA

Majdzik P, Akielaszek-Witczak A, Seybold L (2016) Design of a predictive fault-tolerant control for the battery assembly station. In: Advanced and intelligent computations in diagnosis and control, Springer International Publishing, pp 223–235

Pelletier Francis, Masson Christian, Tahan Antoine (2016) Wind turbine power curve modelling using artificial neural network. Renew Energy 89:207–214

Pourkiaei SM, Ahmadi MH, Hasheminejad SM (2016) Modeling and experimental verification of a 25 W fabricated PEM fuel cell by parametric and GMDH-type neural network, Mech Ind 17(1):105

Roy U, Majumder M (2016) Results and discussions, In: Vulnerability of watersheds to climate change assessed by neural network and analytical hierarchy process, Springer Singapore, pp 47–86

Saaty TL (1978) Modeling unstructured decision problems—the theory of analytical hierarchies. Math Comput Simul 20:147–158

Shaddel M, Javan DS, Baghernia P (2016) Estimation of hourly global solar irradiation on tilted absorbers from horizontal one using artificial neural network for case study of Mashhad. Renew Sustain Energy Rev 53:59–67

Singh Amritpal, Vats Gaurav, Khanduja Dinesh (2016) Exploring tapping potential of solar energy: prioritization of Indian states. Renew Sustain Energy Rev 58:397–406

Strantzali Eleni, Aravossis Konstantinos (2016) Decision making in renewable energy investments: a review. Renew Sustain Energy Rev 55:885–898

Xu Xiaomin, Niu Dongxiao, Qiu Jinpeng, Meiqiong Wu, Wang Peng, Qian Wangyue, Jin Xiang (2016) Comprehensive evaluation of coordination development for regional power grid and renewable energy power supply based on improved matter element extension and TOPSIS method for sustainability. Sustainability 8(2):143

Vaz AGR, Elsinga B, van Sark WGJHM, Brito MC (2016) An artificial neural network to assess the impact of neighbouring photovoltaic systems in power forecasting in Utrecht, the Netherlands. Renew Energy 85:631–641

Chapter 4
Artificial Neural Network

Abstract Artificial Neural Network (ANN) is a technique which can map a relationship within a non-linearly related variables. The development of the model involves selection of network topology,estimation of network weights and validation of the model output by comparing with the desired. Group Method of Data Handling (GMDH) is a new variant of ANN which uses multiple algorithms to find the optimal value of the network weights. The present investigation uses GMDH as a predictive model to estimate the indicator value with the help of input parameters.

Keywords Artificial neural network · Group method of data handling · Predictive model

The Artificial Neural Network (ANN) is widely popular for its flexibility of use, ability to map nonlinearly related variables and reliability of the predictions found from the models developed with the help of the technique.

There are four parameters which are required to be estimated during the development of ANN Models.

The parameters are type of activation function between input and hidden and hidden and output, number of hidden layers, the value of the weights of the connections between input, hidden and output layers.

The first two parameters were generally determined by trial and error method or some search algorithms like Genetic Algorithm (Saemi et al. 2007) etc. and the last two is estimated by different learning algorithms like Back Propagation (Vogl et al. 1988), Levenberg Marquadart (Moré 1978) etc.

In recent years some new variants of neural network was developed which are applied to solve various complex problems which cannot be done with the old ANN methods. Group Method of Data Handling (Ivakhnenko 1970) is an example of such new methods. GMDH has already been utilized in "data mining, knowledge

M. Majumder and A.K. Saha, *Feasibility Model of Solar Energy Plants by ANN and MCDM Techniques*, SpringerBriefs in Energy, DOI 10.1007/978-981-287-308-8_4

discovery (Voss and Feng 2002), prediction (Farlow 1984), complex systems modelling (Madala and Ivaknencho 1994), optimization (Turkson et al. 2016) and pattern recognition".

4.1   Selection of Network Topology

In the selection of network topology for the neural network models, various search algorithms or simple trial and error method is used to identify the optimal network architecture. The number of hidden layers generally depends on number of inputs and amount of training data available for learning. Though, no specific relation has been established till now.

4.2   Training the Model

The value of the weights of the network connections between input, hidden and output layer is also required to be estimated.

In this aspect, various algorithms were utilized to find the weights at which optimal performance can be achieved from a network.

4.3   Testing the Model

The ANN models are trained with part of a dataset where both input and output is known.

After the training of the model is completed based on the "Stop Training Conditions" which can be the maximum number of iterations or level of accuracy as desired by the developer of the model remaining part of the same data set is utilized to "test" the model performance.

Based on the performance result the reliability of the model is estimated.

Sometimes another part of the same dataset is left for validation and once model is trained and validated the remaining part of the data is used for testing the model.

4.4   Applications

The applications of the ANN models are delineated in Fig. 4.1.

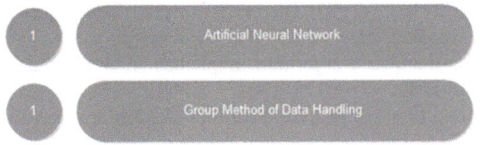

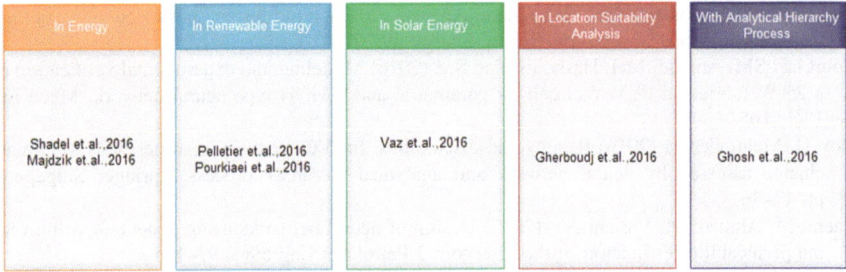

Fig. 4.1 Figure showing various applications of ANN and GMDH

References

Akash Bilal A, Mamlook Rustom, Mohsen Mousa S (1999) Multi-criteria selection of electric power plants using analytical hierarchy process. Electr Power Syst Res 52(1):29–35

Cables E, Lamata MT, Verdegay JL (2016) RIM-reference ideal method in multicriteria decision making. Inf Sci 337:1–10

Capilla JJ, Carrión JA, Alameda-Hernandez E (2016) Optimal site selection for upper reservoirs in pump-back systems, using geographical information systems and multicriteria analysis. Renew Energy 86:429–440

Dong L, Liang H, Gao Z, Luo X, Ren J (2016) Spatial distribution of China? s renewable energy industry: regional features and implications for a harmonious development future, Renew Sustain Energy Rev

Farlow SJ (1984) Self-organizing methods in modelling: GMDH type algorithms. New-York, Bazel: Marcel Decker Inc., p 350

Gherboudj Imen, Ghedira Hosni (2016) Assessment of solar energy potential over the United Arab Emirates using remote sensing and weather forecast data. Renew Sustain Energy Rev 55:1210–1224

Ghosh Soumya, Chakraborty Tilottama, Saha Satyabrata, Majumder Mrinmoy, Pal Manish (2016) Development of the location suitability index for wave energy production by ANN and MCDM techniques. Renew Sustain Energy Rev 59:1017–1028

Ishizaka Alessio, Siraj Sajid, Nemery Philippe (2016) Which energy mix for the UK (United Kingdom)? An evolutive descriptive mapping with the integrated GAIA (graphical analysis for interactive aid)–AHP (analytic hierarchy process) visualization tool. Energy 95:602–611

Ivakhnenko AG (1970) Heuristic Self-Organization in Problems of Engineering Cybernetics. Automatica 6:207–219

Lanjewar P, Rao R, Kale A, Taler J, Oclon P (2016) Evaluation and selection of energy technologies using an integrated graph theory and analytic hierarchy process methods. Decision Sci Lett 5(2):237–348

Linstone Harold A, Turoff Murray (eds) (1975) The Delphi method: Techniques and Applications, vol 29. Addison-Wesley, Reading, MA

Moré JJ (1978) The levenberg-marquardt algorithm: implementation and theory. In: Numerical analysis, Springer Berlin Heidelberg, pp 105–116

Madala HR, Ivakhnenko AG (1994) Inductive Learning Algorithms for Complex Systems Modeling. CRC Press, Boca Raton

Majdzik P, Akielaszek-Witczak A, Seybold L (2016) Design of a predictive fault-tolerant control for the battery assembly station. In: Advanced and intelligent computations in diagnosis and control, Springer International Publishing, pp 223–235

Pelletier Francis, Masson Christian, Tahan Antoine (2016) Wind turbine power curve modelling using artificial neural network. Renew Energy 89:207–214

Pourkiaei SM, Ahmadi MH, Hasheminejad SM (2016) Modeling and experimental verification of a 25 W fabricated PEM fuel cell by parametric and GMDH-type neural network. Mech Ind 17(1):105

Roy U, Majumder M (2016) Results and discussions. In: Vulnerability of watersheds to climate change assessed by neural network and analytical hierarchy process, Springer Singapore, pp 47–86

Saemi M, Ahmadi M, Varjani AY (2007) Design of neural networks using genetic algorithm for the permeability estimation of the reservoir. J Petrol Sci Eng 59(1):97–105

Shaddel M, Javan DS, Baghernia P (2016) Estimation of hourly global solar irradiation on tilted absorbers from horizontal one using artificial neural network for case study of mashhad. Renew Sustain Energy Rev 53:59–67

Singh Amritpal, Vats Gaurav, Khanduja Dinesh (2016) Exploring tapping potential of solar energy: prioritization of Indian states. Renew Sustain Energy Rev 58:397–406

Strantzali Eleni, Aravossis Konstantinos (2016) Decision making in renewable energy investments: a review. Renew Sustain Energy Rev 55:885–898

Turkson RF, Yan F, Ali MKH, Liu B, Hu J (2016) Modeling and multi-objective optimization of engine performance and hydrocarbon emissions via the use of a computer aided engineering code and the NSGA-II genetic algorithm. Sustainability 8(1):72

Xu Xiaomin, Niu Dongxiao, Qiu Jinpeng, Meiqiong Wu, Wang Peng, Qian Wangyue, Jin Xiang (2016) Comprehensive evaluation of coordination development for regional power grid and renewable energy power supply based on improved matter element extension and TOPSIS method for sustainability. Sustainability 8(2):143

Vaz AGR, Elsinga B, van Sark WGJHM, Brito MC (2016) An artificial neural network to assess the impact of neighbouring photovoltaic systems in power forecasting in Utrecht, the Netherlands. Renew Energy 85:631–641

Vogl TP, Mangis JK, Rigler AK, Zink WT, Alkon DL (1988) Accelerating the convergence of the back-propagation method. Biol cybern 59(4–5):257–263

Voss MS, Feng X (2002) A new methodology for emergent system identification using particle swarm optimization (PSO) and the group method data handling (GMDH). In: GECCO, pp 1227–1232

Chapter 5
Methodology

Abstract The present investigation uses Analytical Hierarchy Process and Analytical Network Process to estimate the priority value of the parameters. The ratio of priority value with the value of the beneficiary and non-beneficiary parameters constituted the indicator value which is made directly proportional to the suitability of site to be used for solar power production. The Group Method of Data Handling was used to map the input parameters with the indicator value. Sensitivity analysis was performed and the indicator was used to find the suitability of sites in twelve different locations.

Keywords Indicator · Analytical network process · Sensitivity analysis

5.1 Development of the Model

The model was developed in two steps.

The application of MCDM method was the first step followed by the utilization of GMDH, a new variant of ANN method was the last step.

The MCDM model was used to identify and estimate the priority and priority value of the parameters and GMDH was utilised to estimate the value of the feasibility indicator from the selected parameters.

The indicator was developed with the help of the Eq. 5.1. The indicator was made a direct function of all the parameters which increase the electability of the location for solar power plant installation and inverse function of the parameters which can decrease the suitability.

$$\text{Indicator} = s - \text{Value} = \sum (w_m \times b_m) / \sum (w_n \times b'_n) \qquad (5.1)$$

© The Author(s) 2016
M. Majumder and A.K. Saha, *Feasibility Model of Solar Energy Plants by ANN and MCDM Techniques*, SpringerBriefs in Energy, DOI 10.1007/978-981-287-308-8_5

where w, b and b′ are the priority value, beneficiary and non-beneficiary parameters with respect to the decision objective and m, n are the number of beneficiary and non-beneficiary parameters.

The GMDH model was developed with the selected priority parameters as the input and the indicator or S-Value as the output. The priority value of the parameters was retrieved from both AHP and ANP results. The transformation of the data of the output was performed by Arc Tangent function and model was trained with both GMDH and Combinatorial training algorithm.

In total 24 models were developed among these, 12 models use the input from AHP method as the priority values and the rest 12 utilizes the result of the ANP method as the weights of importance of the input variables.

The neuron or reference function was also changed in between linear and non-linear so that the impact of MCDMs, data transformations, training algorithms and reference function can be incorporated and compared with each other.

Table 5.1 Table showing the characteristic of the 24 models developed for the present investigation

Model no.	No of input	No of output	MCDM	Data transformation	Training	Neuron function
AHNG1	9	1	AHP	None	GMDH	L
#AHNG2	9	1	AHP	None	GMDH	NL
AHAOG3	9	1	AHP	Arc tan of output	GMDH	L
#AHAOG4	9	1	AHP	Arc tan of output	GMDH	NL
AHAIG5	9	1	AHP	Arc tan of input	GMDH	L
#AHAIG6	9	1	AHP	Arc tan of input	GMDH	NL
AHNC7	9	1	AHP	None	C	L
#AHNC8	9	1	AHP	None	C	NL
AHAOC9	9	1	AHP	Arc tan of output	C	L
#AHAOC10	9	1	AHP	Arc tan of output	C	NL
AHAIC11	9	1	AHP	Arc tan of input	C	L
#AHAIC12	9	1	AHP	Arc tan of input	C	NL
ANNG1	9	1	ANP	None	GMDH	L
#ANNG2	9	1	ANP	None	GMDH	NL
ANAOG3	9	1	ANP	Arc tan of output	GMDH	L
#ANAOG4	9	1	ANP	Arc tan of output	GMDH	NL
ANAIG5	9	1	ANP	Arc tan of input	GMDH	L
#ANAIG6	9	1	ANP	Arc tan of input	GMDH	NL
ANNC7	9	1	ANP	None	C	L
#ANNC8	9	1	ANP	None	C	NL
ANAOC9	9	1	ANP	Arc tan of output	C	L
#ANAOC10	9	1	ANP	Arc tan of output	C	NL
ANAIC11	9	1	ANP	Arc tan of input	C	L
#ANAIC12	9	1	ANP	Arc tan of input	C	NL

Table 5.2 Table showing the value of the input parameters for the selected locations

	Agartala	Kolkata	Paris	Berlin	New York	Mexico city	Buenos Aires	Rio de Janeiro	Cairo	Nairobi	Perth	Auckland
Irradiation (B)	0.445	0.445	0.48	0.27	0.39	0.546	0.465	0.517	0.535	0.346	0.556	0.423
No. of consumers (B) (lakhs)	0.5	4.57	2.24	3.5	8.49	21.2	2.965	6.32	6.7	3.138	1.834	1.377
Distance from grid (NB) (km)	12	11	9	9	6	4	9	5	4	3	11	9
Wind speed (NB) (m/s)	2.5	2.2	4.1	6.36	5.77	3.74	4.47	4.4	8	4.76	7.16	7.22
Humidity (NB)(%)	70	65	80	75	62	87	71.4	79.1	59.2	72.8	59	75.9
Sunlight hours (B) (hrs)	12	10	8	7	9	10.5	14	13	10	12	14	14
Cost of labours (NB) (Higher rank to lower cost)	3	3	9	10	1	6	8	7	5	2	12	11
Installation cost (NB) (Higher rank to lower cost)	2	1	6	8	12	11	9	6	10	3	3	3
Rate of energy (B) (Higher rank to higher rate)	11	9	4	2	8	10	12	7	1	6	3	5

The performance metrics like Mean Absolute Error (MAE) and Correlation (r) was used to identify the three better performing models among all the twenty four models considered for the present study. Among the selected three models best model was identified by Root Mean Square Error (RMSE), Mean Relative Error (MRE), Standard Deviation (StDev) and Covariance (Cov). An Equivalent Performance Metrics was developed in this regard. For the first selection EPM was made equal to the ratio of Correlation and MAE. Sixty percent weightage was given to testing and 40 % to training EPM. The second selection was made with the help of Overall Performance Metrics(OPM) which was made equal to inverse of RMSE, MRE, StDev and Cov. Here also 60 % weightage was provided to testing OPM and 40 % to training OPM. Table 5.1 depicts the characteristics of the models developed in this aspect.

5.2   Validation of the Model

The Sensitivity Analysis of the selected model was performed by Single Shot Sensitivity Method.

In total twelve locations were selected from around the world for the feasibility analysis by the new method for installation of solar energy plants. The description of the locations is given in Tables 5.1 and 5.2.

The entire methodology is depicted in Fig. 5.1.

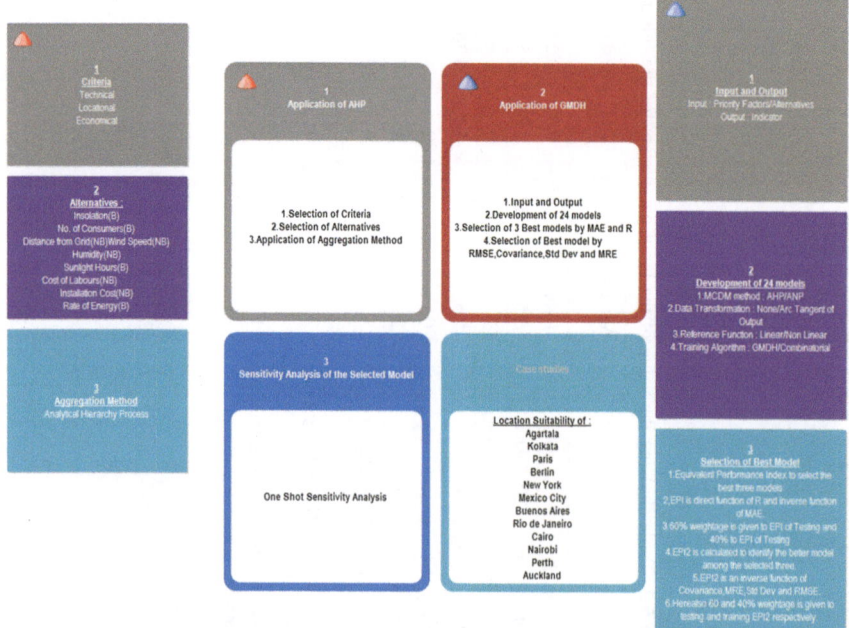

Fig. 5.1 The entire methodology

Chapter 6
Results and Discussions

Abstract The results from the MCDM delineated the importance of insolation parameter in site selection methodologies. The GMDH model trained with GMDH algorithm, output data transformed by Arc tangenet function and priority value retrieved from Analytical Network Process method was found to have the highest performance among the twenty four models developed for the study. The city of New York was found to have the most and Paris have the least suitability for installation of solar power plants among the twelve cities considered for the study.

Keywords New York · Insolation · Arc tangent function

The priority value of the parameters, the most optimal model for prediction of the indicator along with the location with highest feasibility among the 12 selected cities was discussed in the following section.

6.1 Results

The results from the MCDM method are shown in Table 6.1 and 6.2.

The performance metrics of the twenty four predictive models developed for the present study is depicted in Table 6.3 and the metrics of the three better models is depicted in Table 6.4.

The result from the sensitivity analysis of the selected model is depicted in Table 6.5.

The value of the indicators for each of the locations selected for the present investigation is shown in Table 6.6

Figures 6.1, 6.3 and 6.5 shows the comparison of model prediction and actual output derived from the three top ranked models which are model no. #ANAOG4,

© The Author(s) 2016
M. Majumder and A.K. Saha, *Feasibility Model of Solar Energy Plants by ANN and MCDM Techniques*, SpringerBriefs in Energy, DOI 10.1007/978-981-287-308-8_6

Table 6.1 Table showing the results from the AHP MCDM method (B and NB respectively indicates beneficiary and non-beneficiary parameters with respect to study objective)

Criteria	Technical	Locational	Economical		
Criteria weight	0.545	0.182	0.273	Equivalent weight	Rank of the alternative
Insolation (B)	0.352	0.116	0.044	0.225	1
No. of consumers (B)	0.050	0.174	0.152	0.101	5
Distance from grid (NB)	0.117	0.347	0.061	0.144	2
Wind speed (NB)	0.088	0.050	0.038	0.067	8
Humidity (NB)	0.070	0.050	0.034	0.057	9
Sunlight hours (B)	0.176	0.087	0.061	0.128	3
Cost of labours (NB)	0.044	0.069	0.152	0.078	6
Installation cost (NB)	0.044	0.039	0.152	0.073	7
Rate of energy (B)	0.059	0.069	0.305	0.128	4

Table 6.2 Table showing the results from the ANP MCDM method

Criteria	Technical	Locational	Economical		
Criteria weight	0.390	0.327	0.283	Equivalent weight	Rank of the alternative
Insolation (B)	0.352	0.116	0.044	0.187	1
No. of consumers (B)	0.050	0.174	0.152	0.120	4
Distance from grid (NB)	0.117	0.347	0.061	0.177	2
Wind speed (NB)	0.088	0.050	0.038	0.061	8
Humidity (NB)	0.070	0.050	0.034	0.053	9
Sunlight hours (B)	0.176	0.087	0.061	0.114	5
Cost of labours (NB)	0.044	0.069	0.152	0.083	6
Installation cost (NB)	0.044	0.039	0.152	0.073	7
Rate of energy (B)	0.059	0.069	0.305	0.132	3

#AHAOG4 and #AHAIG6. Figures 6.2, 6.4 and 6.5 depicts the distribution of error between the actual and desired output as derived from the top three models (Fig. 6.6).

Equations 6.1, 6.2 and 6.3 respectively represents the model equation of the top three models having the highest EPM among all the twenty four models.

Table 6.3 Table showing the performance metrics of the models developed for the present investigations

Model No.	No of input	No of output	MCDM	Data transformation	Training	Neuron function	Mean absolute error training (%)	Mean absolute error testing (%)	Correlation training (%)	Correlation testing (%)	EPM	Rank
AHNG1	9	1	AHP	None	GMDH	L	17.36	16.01	87.52	94.51	5.541631	17
#AHNG2	9	1	AHP	None	GMDH	NL	2.12	2.62	99.91	99.74	41.24298	4
AHAOG3	9	1	AHP	Arc tan of output	GMDH	L	5.43	4.23	97.61	98.29	20.81062	10
#AHAOG4	9	1	AHP	Arc tan of output	GMDH	NL	1.55	2.15	99.81	99.65	52.20628	2
AHAIG5	9	1	AHP	Arc tan of input	GMDH	L	17.25	15.11	88.29	94.75	5.772642	14
#AHAIG6	9	1	AHP	Arc tan of input	GMDH	NL	1.78	2.29	99.94	99.73	47.84947	3
AHNC7	9	1	AHP	None	C	L	17.14	16.21	87.55	94.42	5.528404	19
#AHNC8	9	1	AHP	None	C	NL	17.43	16.55	87.09	90.64	5.278665	23
AHAOC9	9	1	AHP	Arc tan of output	C	L	5.27	4.13	97.64	98.39	21.38901	9
#AHAOC10	9	1	AHP	Arc tan of output	C	NL	9.14	9.71	92.19	92.58	9.747311	12
AHAIC11	9	1	AHP	Arc tan of input	C	L	17.03	15.19	88.32	94.79	5.789401	13
#AHAIC12	9	1	AHP	Arc tan of input	C	NL	16.88	15.33	88.34	92.04	5.677743	15
ANNG1	9	1	ANP	None	GMDH	L	16.75	17.51	89.38	93.02	5.321632	21

(continued)

Table 6.3 (continued)

Model No.	No of input	No of output	MCDM	Data transformation	Training	Neuron function	Mean absolute error training (%)	Mean absolute error testing (%)	Correlation training (%)	Correlation testing (%)	EPM	Rank
#ANNG2	9	1	ANP	None	GMDH	NL	2.22	3.11	99.88	99.55	36.19535	5
ANAOG3	9	1	ANP	Arc tan of output	GMDH	L	4.16	3.94	97.66	98.18	24.32274	8
#ANAOG4	9	1	ANP	Arc tan of output	GMDH	NL	0.90	1.24	99.88	99.81	90.43297	1
ANAIG5	9	1	ANP	Arc tan of input	GMDH	L	16.91	16.55	89.95	93.98	5.533006	18
#ANAIG6	9	1	ANP	Arc tan of input	GMDH	NL	2.76	3.00	99.81	99.63	34.33264	6
ANNC7	9	1	ANP	None	C	L	16.77	17.68	89.38	92.99	5.286787	22
#ANNC8	9	1	ANP	None	C	NL	16.68	16.88	88.71	90.04	5.327857	20
ANAOC9	9	1	ANP	Arc tan of output	C	L	4.09	3.82	97.67	98.27	24.95672	7
#ANAOC10	9	1	ANP	Arc tan of output	C	NL	6.89	6.53	93.14	94.56	14.08331	11
ANAIC11	9	1	ANP	Arc tan of input	C	L	16.72	16.52	89.99	93.92	5.563133	16
#ANAIC12	9	1	ANP	Arc tan of input	C	NL	20.09	19.56	82.87	87.39	4.328444	24

Table 6.4 Table showing the OPM of the three better models selected from the twenty four models by EPM

Model No.	#ANAOG4		#AHAOG4		#AHAIG6	
	Training	Testing	Training	Testing	Training	Testing
MRE (%)	−5.9E-05	−0.00023	−3.9E-05	−0.00302	0.000617	−0.00474
RMSE (%)	1.26	3.13	2.26	4.78	2.36	3.19
Covar	0.043857	0.042732	0.045951	0.039359	0.292361	0.168009
STDEV	1.26	3.13	2.26	4.77	2.36	3.16
OPM	0.246349131		0.15211249		0.178154816	
Rank	1		3		2	

Table 6.5 Table shows the sensitivity of the alternatives

Alternative	Sensitivity	Rank as per sensitivity
Irradiation (B)	0.885	1
No. of consumers (B)	0.679	2
Distance from grid (NB)	−0.896	9
Wind speed (NB)	−0.241	6
Humidity (NB)	−0.228	5
Sunlight hours (B)	0.555	4
Cost of labours (NB)	−0.465	8
Installation cost (NB)	−0.430	7
Rate of energy (B)	0.667	3

Table 6.6 Table showing the rank of the location compared to the other selected location based on the S value of the location

Location	S-value	Rank as per S-value
Agartala	0.375	6
Kolkata	0.378	4
Paris	0.352	12
Berlin	0.352	9
New York	0.669	1
Mexico city	0.378	5
Buenos Aires	0.353	8
Rio de Janeiro	0.358	7
Cairo	0.404	2
Nairobi	0.396	3
Perth	0.352	11
Auckland	0.352	10

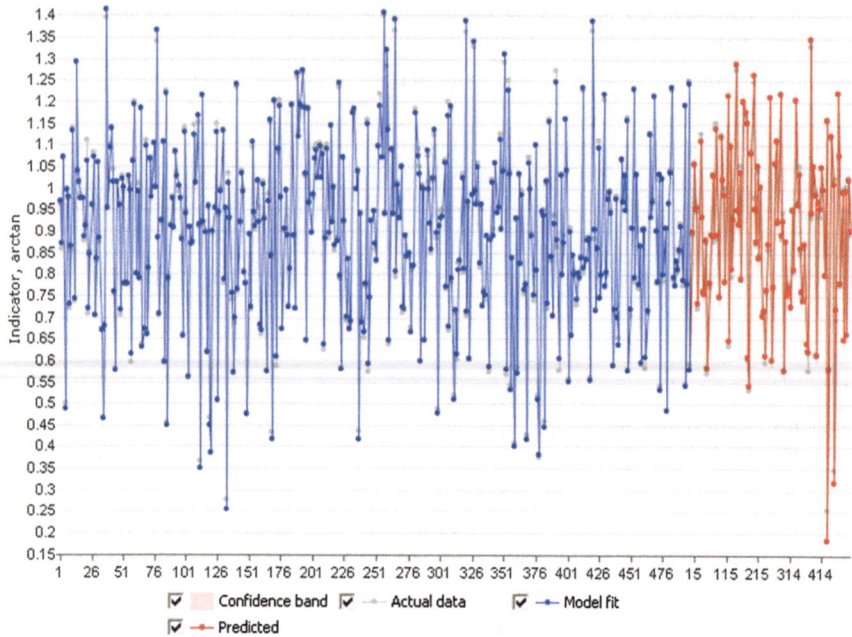

Fig. 6.1 Figure showing the observed and predicted output for the Model No. #ANAOG4

6.2 Discussions

The results from AHP as well as ANP method depicted that Insolation of the location and humidity was respectively found to be the most and least important parameter among all the compared factors with respect to the decision goal. The importance of insolation in solar energy production is well documented in papers like Khabibrakhmanov et al. (2016), Stökler et al. (2016) etc.

The model, which used the output of ANP, data of output was transformed by Arc Tangent function and the neural function was made non-linear; was found to have the best performance metric among all the twenty four models developed for the present investigation. The reliability of ANP is more than AHP as the former method was developed to compensate the drawbacks of the latter (Mirzaei and Avakhdarestani 2016).

The sensitivity of the most important parameter was found to be most sensitive among all the parameters but the sensitivity of least important parameter was not equal to the least sensitive parameter which was found to be distance of the location from the consumers or grid. Humidity which was the least important parameter

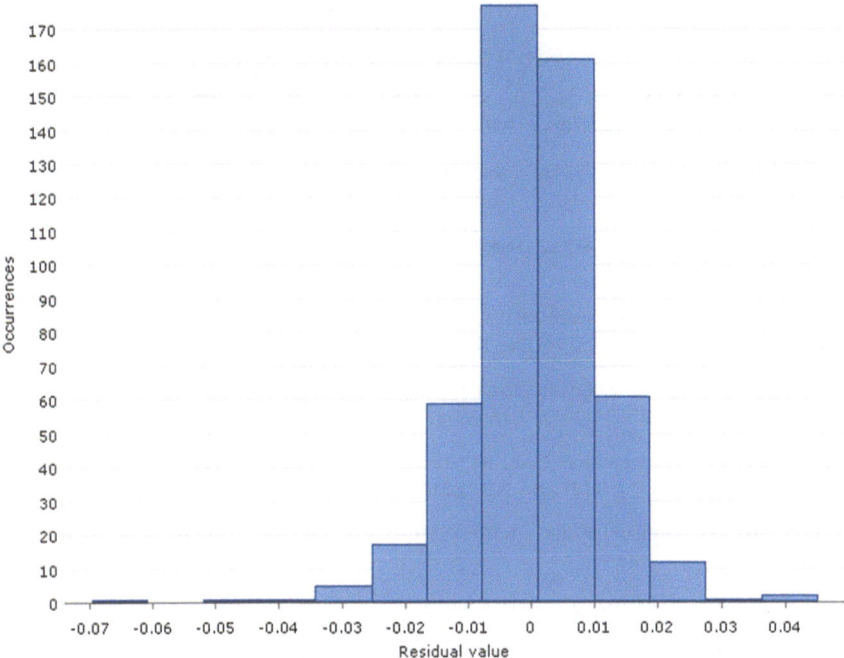

Fig. 6.2 Figure showing the distribution of residuals from predicted and observed output for model no. #ANAOG4

Eq. 6.1. Equation for Model No. #ANAOG4

$Y1 = 0.00156743 + N259*N2*1.21617 - N259^2*0.599898 + N2*1.00084 - N2^2*0.61722$

$N2 = 0.12696 - N529*0.292544 + N529^2*0.166203 + N3*0.999632$

$N3 = -0.00787956 + N428*0.0770172 + N428*N4*0.322956 - N428^2*0.206227 + N4*0.947159 - N4^2*0.131107$

$N4 = -0.0212613 + $ "Cost of Labours(NB)"$*0.0561824 - $ "Cost of Labours(NB)"$*N5*0.0338796 - $ "Cost of Labours(NB)"$^2*0.0227789 + N5*1.01722$

$N5 = -0.0382937 + N443*0.173514 + N443*N6*0.18961 - N443^2*0.193436 + N6*0.918819 - N6^2*0.0487254$

$N6 = 0.00489685 - $ "Installation Cost(NB)"$*0.0408454 + $ "Installation Cost(NB)"$^2*0.0416524 + N7*1.00149$

N7 = 0.282083 - N575*0.704387 - N575*N8*0.0706177 + N575^2*0.430953 + N8*1.0639

N8 = 0.176227 - N541*0.452119 + N541^2*0.255286 + N9*1.04787 - N9^2*0.0265818

N9   =   -0.00382242   -   Humidity(NB)*0.024319   -   Humidity(NB)*N10*0.0263933   + Humidity(NB)^2*0.0448926 + N10*1.01396

N10 = -0.00157438 - N420*0.13349 - N420*N11*0.573069 + N420^2*0.36443 + N11*1.12915 + N11^2*0.211904

N11 = 0.0206975 - N558*0.280952 - N558*N12*0.430566 + N558^2*0.393795 + N12*1.20086 + N12^2*0.0955609

N12 = 0.0321451 - "Wind Speed(NB)"*0.14029 + "Wind Speed(NB)"*N14*0.0456451 + "Wind Speed(NB)"^2*0.087892 + N14*0.997198 - N14^2*0.0102148

N14 = -0.0123614 + "No. of Consumers(B)"*0.0892355 - "No. of Consumers(B)"*N15*0.0624564 - "No. of Consumers(B)"^2*0.0480702 + N15*0.988324 + N15^2*0.0273674

N15 = -0.155377 + "Distance from Grid(NB)"*0.260241 - "Distance from Grid(NB)"*N21*0.175093 - "Distance from Grid(NB)"^2*0.121126 + N21*1.25099 - N21^2*0.102087

N21 = -0.907638 + N586*1.59429 - N586*N28*0.615027 - N586^2*0.530762 + N28*1.36029 + N28^2*0.097369

N28 = -0.00200891 + N45*0.996576 - N45*N50*0.509831 + N50^2*0.515504

N50 = -0.00948921 + N75*0.831036 - N75*N77*0.344539 + N77*0.180128 + N77^2*0.343608

N77 = 0.0446024 + Insolation(B)*0.225348 - Insolation(B)*N138*0.572327 + Insolation(B)^2*0.280577 + N138*0.729493 + N138^2*0.311127

N138 = -0.0440592 - "Wind Speed(NB)"*N269*0.148021 + N269*1.20256 - N269^2*0.0829274

N269 = -2.38149 + N466*2.10281 - N466*N540*0.507417 - N466^2*0.333835 + N540*3.1111 - N540^2*0.874154

N540 = -0.427288 + N580^2*0.553377 + N586*0.975686

N580 = 0.849409 + "No. of Consumers(B)"*0.212537 - "Installation Cost(NB)"^2*0.173446

N466 = 0.0262429 + N506*N541*1.08129

N506 = 1.13898 - "Distance from Grid(NB)"*0.542643 + "Distance from Grid(NB)"^2*0.184605 - Humidity(NB)^2*0.138635

N75 = 0.953388 - N495*1.49297 + N495*N207*1.68713 + N207*0.36062 - N207^2*0.4992

N207 = 0.93185 - N593*2.88689 + N593^2*2.03972 + N345*1.04508 - N345^2*0.0402777

N345 = -1.26066 + N476*1.55686 - N476*N499*0.612868 + N499*1.23427 + N499^2*0.179174

N499 = -3.28066 + N573*3.61307 - N573*N579*2.94428 + N579*3.68497

N579 = 0.686337 + "No. of Consumers(B)"*0.231102 + "Sunlight Hours(B)"*0.192479

N573 = 0.894938 - "Cost of Labours(NB)"*0.21799 + "Rate of Energy(B)"*0.210632

N476 = 0.605007 - "Distance from Grid(NB)"*0.362452 + N538^2*0.56386

N593 = 0.998498 + "Wind Speed(NB)"*0.0484993 - "Wind Speed(NB)"^2*0.224527 - Humidity(NB)^2*0.151062

N495 = 1.20832 - "Distance from Grid(NB)"*0.583515 + "Distance from Grid(NB)"*"Cost of Labours(NB)"*0.0757884 + "Distance from Grid(NB)"^2*0.195343 - "Cost of Labours(NB)"*0.235328

N45 = -0.0247527 - Humidity(NB)*N78*0.139122 + Humidity(NB)^2*0.0469846 + N78*1.09385 - N78^2*0.0173468

N78 = 0.0716159 + Insolation(B)*0.194368 - Insolation(B)*N188*0.659598 + Insolation(B)^2*0.350679 + N188*0.659431 + N188^2*0.394286

N188 = -0.216079 + N276*0.319137 + N276^2*0.257689 + N409*1.12616 - N409^2*0.477889

N409 = -0.271454 + N559^2*0.542854 + N477*0.620291 + N477^2*0.203284

N477 = -3.51285 + N501*2.63917 - N501*N575*1.70698 + N575*4.92192 - N575^2*1.23419

N501 = 1.01187 - "Distance from Grid(NB)"*0.554541 + "Distance from Grid(NB)"^2*0.188423 + "Sunlight Hours(B)"*0.179375

N559 = 0.749635 + Insolation(B)*0.554752 + Insolation(B)*Humidity(NB)*0.0900062 - Insolation(B)^2*0.2671 - Humidity(NB)^2*0.182582

N276 = -1.30611 + N455*1.49351 - N455*N500*0.566422 + N500*1.47059

N500 = -3.61209 + N574*3.95093 - N574*N586*3.2142 + N586*3.95713

N455 = 0.0203682 + N502*N538*1.08885

N586 = 0.70382 + "Sunlight Hours(B)"*0.213801 - "Sunlight Hours(B)"^2*0.0515114 + "Rate of Energy(B)"*0.204168

N558 = 0.610279 + Insolation(B)*0.618598 - Insolation(B)*"Sunlight Hours(B)"*0.0427278 - Insolation(B)^2*0.271001 + "Sunlight Hours(B)"*0.170899

N420 = -1.66465 + N488*2.73638 - N488*N575*1.88253 + N575*0.725322 + N575^2*1.19406

N541 = 0.801365 + Insolation(B)*0.552278 - Insolation(B)^2*0.218985 - "Cost of Labours(NB)"*0.210057

N443 = -0.429546 + N488*0.989639 + N574^2*0.540055

N574 = 0.895697 + "No. of Consumers(B)"*0.20836 - "Cost of Labours(NB)"*0.212644

N488 = 0.894782 + Insolation(B)*0.599263 - Insolation(B)^2*0.266597 - "Distance from Grid(NB)"*0.580207 + "Distance from Grid(NB)"^2*0.224027

N428 = -1.34201 + N489*1.50337 - N489*N538*0.56864 + N538*1.50251

N538 = 0.776441 + Insolation(B)*0.657117 - Insolation(B)^2*0.308264 - "Installation Cost(NB)"*0.198054

N489 = 0.99462 - "Distance from Grid(NB)"*0.681734 + "Distance from Grid(NB)"^2*0.299353 + "Rate of Energy(B)"*0.253039

N529 = 1.0914 - N578*2.26671 + N578*N585*3.6303 - N585^2*1.33712

N585 = 0.97881 - "Installation Cost(NB)"*0.327052 + "Installation Cost(NB)"*"Rate of Energy(B)"*0.326388

N259 = 0.521375 - N594*2.4008 + N594^2*1.88334 + N381*1.26707 - N381^2*0.156057

N381 = -0.463711 + N483*0.990978 + N498^2*0.575365

N498 = -4.43738 + N575*6.55237 - N575*N578*2.47817 - N575^2*1.85043 + N578*3.29022

N578 = 0.94542 + "Sunlight Hours(B)"*0.121894 + "Sunlight Hours(B)"*"Cost of Labours(NB)"*0.119603 - "Cost of Labours(NB)"*0.279945

N575 = 0.647318 + "No. of Consumers(B)"*0.290161 - "No. of Consumers(B)"*"Rate of Energy(B)"*0.17107 + "Rate of Energy(B)"*0.287067

N483 = 0.246458 + Insolation(B)*0.604486 - Insolation(B)^2*0.276419 + N502^2*0.544203

N502 = 1.14637 - "Distance from Grid(NB)"*0.539238 + "Distance from Grid(NB)"^2*0.188103 - "Wind Speed(NB)"^2*0.154891

N594 = 0.996285 - Humidity(NB)^2*0.135367 - "Installation Cost(NB)"^2*0.163026

among the nine factors considered in the study was found to be 5th most sensitive parameter. All the non-beneficiary parameter was found to have a negative sensitivity which is coherent to the study objective.

The comparison among the selected location yielded that New York and Paris has the most and least feasibility for installation of solar power plant among the 12 cities compared in the study. Solar power in the United States means "utility-scale solar power plants as well as local distributed generation". Roof top installations of solar panels are the biggest contributor. The capacity of solar power is increasing rapidly in recent years as costs have declined. USA now has 20 GW of installed solar capacity (18.3 GW solar PV; 1.7 GW concentrated solar power) in 2014 (Munsell 2015).Eight among the ten largest solar power plant is situated in America and nearly 6,45,000 household and business entities has installed solar power plants to satisfy their need of energy(SEIA 2015).

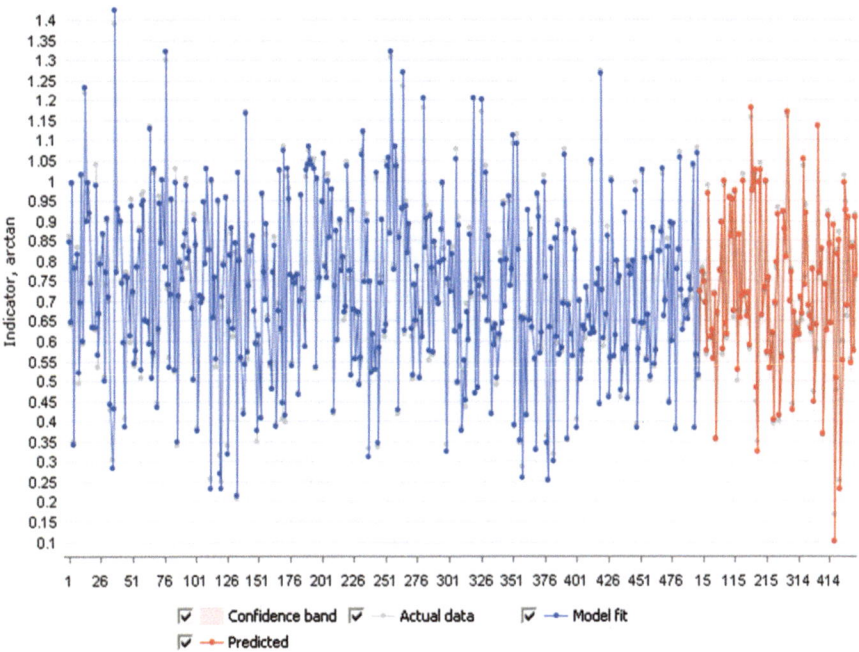

Fig. 6.3 Figure showing the comparison between observed and predicted output for model no. #AHAOG4

6.3   Scientific Benefit

The result of the model shows that this method can be applied for pre-installation survey of solar power plant which can reduce the cost of such surveys manifold. Not only this, the method can prevent unnecessary wastage of fund by selecting the most feasible locations from the data collected from the area of interest.

The model equation can be used to develop stand alone apps which can predict the feasibility of location in situ and can also estimate feasibility of the area of interest in the future if adequate date of the input variables can be fed to the model.

6.4   Model Limitation

Although the model has many benefits but as the model is method dependent the results will be non-uniform when a new method is used to estimate the indicator value. Even the priority values of the parameters may also change if the type of survey is changed.

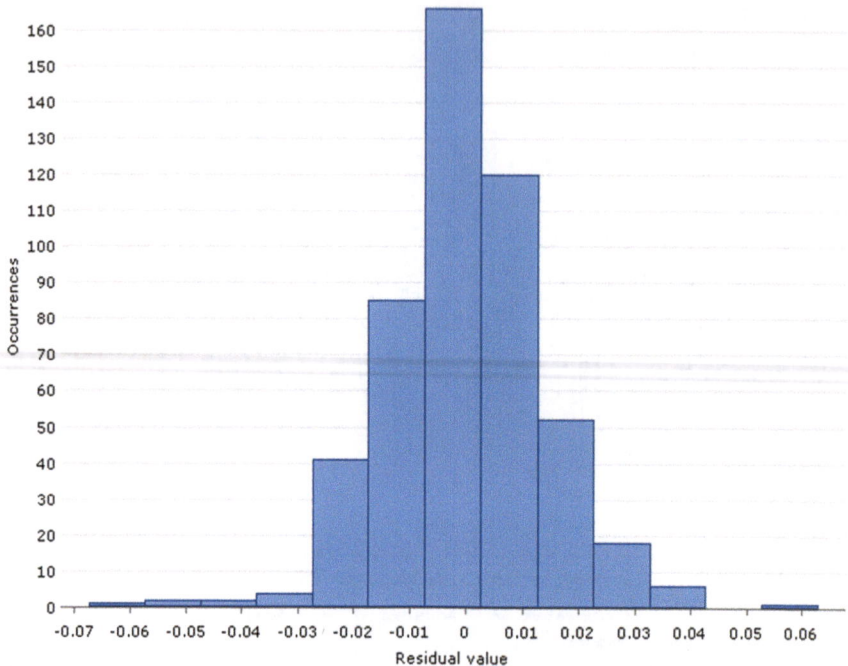

Fig. 6.4 Figure showing the distribution of residuals from observed and predicted output for model no. #AHAOG4

Eq. 6.2. Equation for Model No. #AHAOG4

$Y1 = -0.00432303 - N258*N2*1.28818 + N258^2*0.653804 + N2*1.00928 + N2^2*0.627449$

$N2 = -0.00304297 + N301*0.0725253 + N301*N3*0.518535 - N301^2*0.338302 + N3*0.94221 - N3^2*0.192472$

$N3 = -0.124814 + N554*0.355851 - N554^2*0.242015 + N4*0.987142 + N4^2*0.0100354$

$N4 = 0.00283066 + N125*N5*0.149448 - N125^2*0.146958 + N5*0.994145$

$N5 = 0.00735958 + N363*0.0337014 + N363*N6*1.2568 - N363^2*0.654262 + N6*0.956292 - N6^2*0.596372$

$N6 = -0.0258012 - $ "Installation Cost(NB)"$*0.0182107 - $ "Installation Cost(NB)"$*N7*0.0460453 + $ "Installation Cost(NB)"$"^2*0.0505478 + N7*1.07497 - N7^2*0.036667$

N7 = -0.0104252 + "Rate of Energy(B)"*0.0795187 - "Rate of Energy(B)"*N8*0.0210801 - "Rate of Energy(B)"^2*0.0525561 + N8*0.983584 + N8^2*0.0140792

N8 = 0.00557137 - Humidity(NB)*0.058197 - Humidity(NB)*N9*0.0132041 + Humidity(NB)^2*0.0665226 + N9*1.00821

N9 = -0.00478927 + N10*1.01256 + N10*N11*0.770807 - N10^2*0.778276

N11 = 0.0113347 - "Wind Speed(NB)"*0.0954297 + "Wind Speed(NB)"^2*0.0848205 + N14*1.02396 - N14^2*0.0162768

N14 = 0.0111247 - N84*0.826981 + N84*N16*9.81526 - N84^2*4.63351 + N16*1.79996 - N16^2*5.16311

N16 = 0.00185402 - N24*N34*7.57108 + N24^2*4.15267 + N34*0.983388 + N34^2*3.43254

N34 = -0.0172277 - N381*N42*2.10321 + N381^2*1.08667 + N42*1.02575 + N42^2*1.00066

N42 = -0.00527082 + N73*0.609123 + N106*0.398066

N73 = -0.0697957 + N355*0.502845 - N355*N125*1.30172 + N355^2*0.452776 + N125*0.663315 + N125^2*0.750101

N355 = -0.501533 + N444*0.771914 + N444*N503*0.239303 + N503*0.736142

N444 = 0.0439289 + N507*N545*1.28039

N545 = 0.88184 + "Sunlight Hours(B)"*"Cost of Labours(NB)"*0.122791 + "Sunlight Hours(B)"^2*0.0756858 - "Cost of Labours(NB)"*0.414913

N381 = -0.537876 + N464*0.475002 + N464^2*0.355178 + N529*0.985858

N529 = 0.227987 - N575*0.761524 + N575*N579*2.50574 - N579^2*0.519211

N575 = 0.903356 - "Distance from Grid(NB)"*0.321256 + "Distance from Grid(NB)"^2*0.118191 - "Wind Speed(NB)"^2*0.160153

N464 = -0.670767 + N519*0.961816 + N548*0.953035

N548 = 0.606306 + Insolation(B)*0.59167 - Insolation(B)^2*0.257386 - Humidity(NB)*0.161972

N84 = -0.020898 + N553*N137*0.277295 + N137*0.902262 - N137^2*0.101141

N137 = -0.0345744 + N260*0.597442 + N277*0.461911 - N277^2*0.0154184

N277 = -0.347636 + N413*0.390423 + N413*N544*0.484602 + N413^2*0.164271 + N544*0.600551

N544 = -0.0617656 + N578^2*0.73604 + N581^2*0.727421

N413 = -0.685286 + N496*0.967532 + N507*0.967122

N496 = 0.772452 - "Cost of Labours(NB)"*0.355191 + "Rate of Energy(B)"*0.259701

N10 = 0.0213606 - "Wind Speed(NB)"*0.106447 + "Wind Speed(NB)"*N13*0.00767467 + "Wind Speed(NB)"^2*0.0929139 + N13*0.997406

N13 = -0.0462971 + "Sunlight Hours(B)"*0.119711 - "Sunlight Hours(B)"*N18*0.139935 + N18*1.03974 + N18^2*0.0170421

N18 = 0.00475744 - N24*N36*4.82949 + N24^2*2.78399 + N36*0.977811 + N36^2*2.06318

N36 = 0.292768 - N499*0.766927 + N499^2*0.458389 + N41*1.02117

N41 = -0.00516434 + N72*0.618909 + N106*0.388135

N106 = 0.555391 - N542*0.990746 + N542*N161*1.32723 + N161*0.480327 - N161^2*0.307238

N161 = -1.00098 + N584*2.11322 - N584^2*1.04249 + N263*1.07346 - N263^2*0.0650153

N263 = 0.108007 + N416*0.153614 + N416*N536*0.861299 + N416^2*0.155478 - N536*0.47563 + N536^2*0.563613

N536 = 0.0280219 + N574*N577*3.01045 - N574^2*0.839729 - N577^2*0.840438

N416 = -1.27601 + N488*1.7286 - N488*N562*0.991258 + N562*1.73832

N584 = 0.715573 - "Wind Speed(NB)"*"Sunlight Hours(B)"*0.1243 - "Wind Speed(NB)"^2*0.115075 + "Sunlight Hours(B)"*0.181893

N72 = -0.0540736 + N125*0.393457 - N125*N214*3.23924 + N125^2*1.79405 + N214*0.738379 + N214^2*1.36448

N214 = -0.00842263 + N272*0.31043 + N272*N289*0.338416 + N289*0.674632 - N289^2*0.303349

N289 = -0.328776 + N441*0.722389 + N441*N503*0.335585 + N503*0.236143 + N503^2*0.322061

N503 = -2.12949 + N564*3.24367 - N564*N579*1.67993 - N564^2*0.624957 + N579*2.35616

N564 = 0.741781 + "No. of Consumers(B)"*0.233971 - "Distance from Grid(NB)"*0.368105 + "Distance from Grid(NB)"^2*0.148772

N441 = -0.181296 + N507^2*0.647789 + N527*0.595049 + N527^2*0.216782

N272 = -1.62856 + N420*1.104 - N420*N534*0.152446 + N534*3.33656 - N534^2*1.47578

N534 = 0.026131 + N572*N579*1.31491

N572 = 0.641865 + "No. of Consumers(B)"*0.333149 - "No. of Consumers(B)"*"Wind Speed(NB)"*0.229354 - "Wind Speed(NB)"^2*0.0539168

N420 = -0.67297 + N488*N558*1.34404 + N558*1.82229 - N558^2*1.19271

N499 = 1.08723 - "Cost of Labours(NB)"*0.539758 + "Cost of Labours(NB)"*"Installation Cost(NB)"*0.398346 - "Installation Cost(NB)"*0.293479 - "Installation Cost(NB)"^2*0.122435

N24 = -0.00412419 - Insolation(B)*0.125828 - Insolation(B)*N54*0.103616 + Insolation(B)^2*0.201067 + N54*1.05436

N54 = 0.547462 - N527*1.4247 + N527*N77*0.339976 + N527^2*0.804828 + N77*0.911137 - N77^2*0.118082

N77 = 0.00555325 + N553*N136*0.351221 - N553^2*0.0419511 + N136*0.835805 - N136^2*0.0927715

N136 = -0.0290081 + N260*0.601256 + N281*0.438308

N281 = -0.58274 + N423*0.673062 + N423^2*0.209795 + N524*0.959431

N524 = -0.0804737 + N574*N579*1.51534

N579 = 0.540465 + "Sunlight Hours(B)"*0.15983 - "Sunlight Hours(B)"^2*0.0470701 + "Rate of Energy(B)"*0.250935

N574 = 0.899659 - "Distance from Grid(NB)"*0.320872 + "Distance from Grid(NB)"^2*0.110694 - Humidity(NB)^2*0.157871

N423 = -0.705884 + N507*0.999352 - N507*N519*0.0442033 + N519*0.995838

N519 = 0.79038 + "No. of Consumers(B)"*0.223476 - "Cost of Labours(NB)"*0.348749

N507 = 0.644973 + Insolation(B)*0.657569 - Insolation(B)^2*0.30687 - "Installation Cost(NB)"*0.264297

N553 = 1.05278 + N581*N582*1.32023 - N582*2.89615 + N582^2*2.01546

N581 = 0.794346 - "Distance from Grid(NB)"*0.345607 + "Distance from Grid(NB)"^2*0.129689 + "Sunlight Hours(B)"*0.125169

N363 = -0.208795 + N462*0.467414 + N462*N480*0.537286 + N480*0.420125

N462 = -1.52647 + N525*2.63001 - N525*N562*2.14672 + N562*1.3547 + N562^2*0.897469

N525 = 0.626062 + Insolation(B)*0.609884 - Insolation(B)^2*0.280342 - "Distance from Grid(NB)"*0.212927

N125 = 0.0237102 - Humidity(NB)*0.0346636 - Humidity(NB)*N260*0.121876 + N260*1.049

N260 = -0.36209 + N480*0.652458 + N480*N485*0.471749 + N485*0.279329 + N485^2*0.283557

N480 = 0.569929 - N542*1.45785 + N542*N558*3.28086 - N558^2*0.975215

N542 = 0.948819 - "Wind Speed(NB)"^2*0.145366 - "Cost of Labours(NB)"*0.337854

N554 = 0.0491571 + N580*N582*1.27161

N582 = 0.847888 - "Wind Speed(NB)"^2*0.176944 - Humidity(NB)^2*0.16754

N580 = 0.636463 + "No. of Consumers(B)"*"Sunlight Hours(B)"*0.247709 + "No. of Consumers(B)"^2*0.112555

N301 = -0.381953 + N472*0.0980291 + N472*N487*0.404169 + N472^2*0.397922 + N487*0.991294 - N487^2*0.223774

N487 = -0.0355902 + N558*N562*1.43122

N558 = 1.03706 - "Distance from Grid(NB)"*0.456151 + "Distance from Grid(NB)"*"Installation Cost(NB)"*0.308559 + "Distance from Grid(NB)"^2*0.0955249 - "Installation Cost(NB)"*0.381722

N472 = -1.31413 + N527*1.83253 - N527*N556*1.15697 + N556*1.80849

N527 = 1.00376 + Humidity(NB)*"Cost of Labours(NB)"*0.363693 - Humidity(NB)^2*0.319121 - "Cost of Labours(NB)"*0.52484

N258 = -0.299582 + N485*0.794498 - N485*N279*1.56706 + N485^2*0.293958 + N279*1.00384 + N279^2*0.728489

N279 = -0.478602 + N411*0.374639 + N411*N535*0.66725 + N411^2*0.0824034 + N535*1.01332 - N535^2*0.385167

N535 = -0.0235494 + N577*N578*1.40815

N578 = 0.671242 + "No. of Consumers(B)"*0.21697 - Humidity(NB)^2*0.151178

N577 = 0.810328 - "Sunlight Hours(B)"*"Installation Cost(NB)"*0.0416061 + "Sunlight Hours(B)"^2*0.134275 - "Installation Cost(NB)"*0.218061

N411 = -0.748444 + N488*1.00716 + N555*1.01363

N555 = 0.737659 - "Distance from Grid(NB)"*0.476985 + "Distance from Grid(NB)"^2*0.239626 + "Rate of Energy(B)"*0.282491

N488 = 0.715395 + Insolation(B)*0.49161 - Insolation(B)^2*0.16175 - "Cost of Labours(NB)"*0.347744

N485 = -2.30128 + N556*3.13082 - N556*N562*2.86831 + N562*3.10999

N562 = 0.449512 + "No. of Consumers(B)"*0.310107 - "No. of Consumers(B)"*"Rate of Energy(B)"*0.18227 + "Rate of Energy(B)"*0.338421

N556 = 0.479775 + Insolation(B)*0.602563 - Insolation(B)^2*0.27654 + "Sunlight Hours(B)"*0.100628

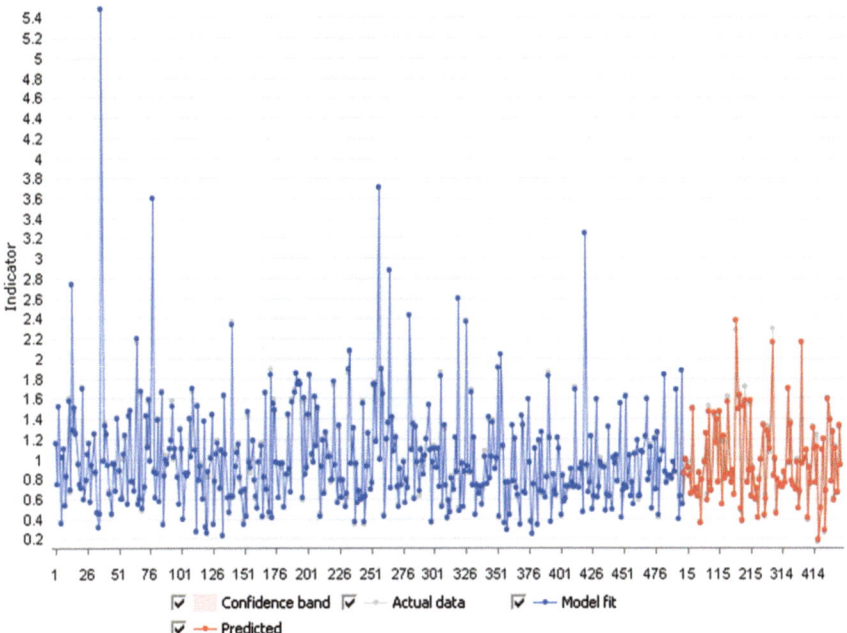

Fig. 6.5 Figure showing the comparison between observed and predicted output for model no. #AHAIG6

If the type of survey is changed new parameters can emerge as more important than the existing parameters. The only solution to this problem is to devise some policy so that the method type can be fixed.

Besides the limitations as described in the previous paragraph, this model can reliably identify suitable locations for solar energy installation and can reduce sufficient amount of expenditures by selecting the optimal location before installation as is eminent from the accuracy of the selected model(around 99.10 %).

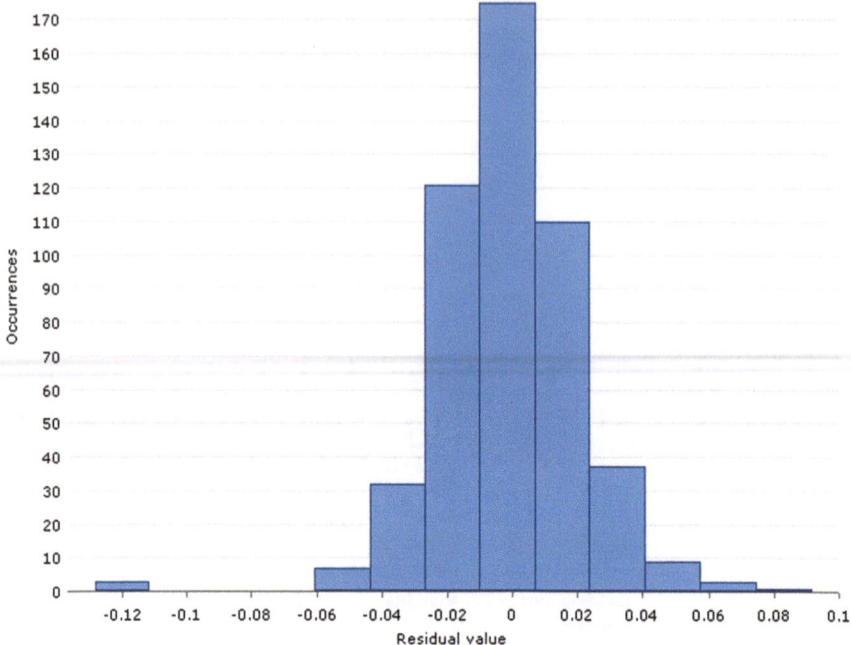

Fig. 6.6 Figure showing the distribution of residuals from observed and predicted output for model no. #AHAIG6

Eq. 6.3. Equation for Model No. #AHAIG6

$Y1 = -0.00225819 + N364*0.0283646 + N364*N2*0.122642 - N364^2*0.0749053 + N2*0.97756 - N2^2*0.0489234$

$N2 = 0.00248107 + N11*N3*20.6146 - N11^2*10.4372 + N3*0.999004 - N3^2*10.1769$

$N3 = 0.000728786 - N41*N4*1.51588 + N41^2*0.769801 + N4*0.996334 + N4^2*0.746534$

$N4 = 0.0064493 - N270*0.0631495 - N270*N5*0.124417 + N270^2*0.0740538 + N5*1.05371 + N5^2*0.0518561$

$N5 = 0.0691041 - N565*0.179783 - N565*N6*0.0179892 + N565^2*0.105557 + N6*1.01858$

$N6 = -0.023442 + N472*0.0656039 + N472*N7*0.0254035 - N472^2*0.0399274 + N7*0.979685 - N7^2*0.00412332$

N7 = -0.00406486 - N546*0.0426509 - N546*N8*0.074137 + N546^2*0.0582791 + N8*1.04967 + N8^2*0.0108036

N8 = -0.00599103 + N336*0.0810524 + N336*N9*0.106363 - N336^2*0.0720035 + N9*0.928063 - N9^2*0.0356649

N9 = 0.0158124 - N508*0.0481981 - N508*N10*0.0201678 + N508^2*0.036145 + N10*1.01072 + N10^2*0.00320095

N10 = -0.000391794 - N22*N11*0.175051 + N11*1.00078 + N11^2*0.174589

N508 = 1.74698 - N580*1.68771 + N580*N585*2.64676 - N585*1.71061

N336 = 0.283769 + N424*N526*1.09801 - N424^2*0.0698311 - N526*0.444645 + N526^2*0.123329

N424 = 0.850918 - N511*0.738502 + N511*N552*1.67457 - N552*0.795991

N546 = 0.169983 + N572*0.0693177 + N572*N583*0.752123

N583 = 0.157158 + N595*N597*0.838003

N595 = 0.88242 - ""Wind Speed(NB)", arctan"*""Sunlight Hours(B)", arctan"*1.18766 + ""Sunlight Hours(B)", arctan"*0.810332

N572 = 0.289513 + N590*N593*0.69936

N590 = 0.70737 + ""No. of Consumers(B)", arctan"*0.53963 - ""No. of Consumers(B)", arctan"*""Humidity(NB), arctan"*0.878693 + ""No. of Consumers(B)", arctan"^2*0.513385 + "Humidity(NB), arctan"*0.759335 - "Humidity(NB), arctan"^2*0.993367

N472 = 1.4611 - N518*1.85752 + N518*N580*2.14068 + N518^2*0.35173 - N580*1.12654

N41 = -0.214189 + N506*0.262871 - N506*N69*0.247283 + N506^2*0.0294906 + N69*1.13516 + N69^2*0.0422642

N69 = -0.00508283 + N106*0.823254 + N106*N109*0.17425 - N106^2*0.173437 + N109*0.181514

N109 = -0.0277366 + N221*1.17612 - N221^2*0.203463 - N264*0.134732 + N264^2*0.193112

N221 = 0.17403 + N298*0.651308 + N355^2*0.139468

N355 = 0.742376 - N459*0.914654 + N459*N542*1.42949 + N459^2*0.20421 - N542*0.499307

N542 = -0.0538872 + N591*N592*1.05319

N592 = 1.02623 - ""Wind Speed(NB)", arctan"*0.358418 - ""Wind Speed(NB)", arctan"*""Rate of Energy(B)", arctan"*0.491912 + ""Rate of Energy(B)", arctan"^2*0.917217

N459 = 0.0515584 + N521*N540*0.943554

N540 = 1.09894 + "Insolation(B), arctan"*0.84915 - ""Installation Cost(NB)", arctan"*1.43719 + ""Installation Cost(NB)", arctan"^2*0.68077

N298 = -0.043649 - N440*0.0734578 + N440*N532*1.11653

N440 = 1.63058 - N509*1.50339 + N509*N574*2.40055 - N574*1.53916

N106 = -0.00765487 + N416*0.242161 + N158*0.715785 + N158^2*0.0387452

N158 = 0.118904 + N309*0.425241 + N309*N323*0.114165 + N323*0.315823

N323 = 0.710042 - N455*0.868684 + N455*N512*1.30726 + N455^2*0.217822 - N512*0.505545 + N512^2*0.0958348

N512 = -0.0285967 + N565*N591*1.02756

N455 = 0.126037 + N518*N552*1.57487 - N518^2*0.334677 - N552^2*0.332229

N506 = 0.669985 + N566*N585*1.00596 - N585*1.34395 + N585^2*0.639305

N11 = 0.0222676 - "Humidity(NB), arctan"*0.106458 + "Humidity(NB), arctan"^2*0.103167 + N12*0.998954

N12 = -0.000250837 + N77*N13*0.890345 - N77^2*0.424882 + N13*1.00135 - N13^2*0.464663

N13 = 0.0301515 - ""Wind Speed(NB)", arctan"*0.154264 + ""Wind Speed(NB)", arctan"*N14*0.018205 + ""Wind Speed(NB)", arctan"^2*0.148269 + N14*0.993639

N14 = -0.00265674 + N416*0.0393828 + N15*0.956909 + N15^2*0.00491699

N15 = 0.0602813 - N439*0.121383 + N439^2*0.0386649 + N16*1.02247 - N16^2*0.00511862

N16 = -0.0320046 + ""No. of Consumers(B)", arctan"*0.211046 - ""No. of Consumers(B)", arctan"^2*0.255312 + N17*1.00136

N17 = 0.0210182 - "Insolation(B), arctan"*0.18001 - "Insolation(B), arctan"*N18*0.0146636 + "Insolation(B), arctan"^2*0.231015 + N18*1.00788

N18 = 0.0356425 - ""Cost of Labours(NB)", arctan"*0.187082 + ""Cost of Labours(NB)", arctan"^2*0.206063 + N19*0.995103

N19 = -0.0579736 + ""Rate of Energy(B)", arctan"*0.340385 - ""Rate of Energy(B)", arctan"^2*0.360855 + N20*0.996783

N20 = 0.0601125 - N534*0.159297 + N534^2*0.103498 + N22*0.987927

N22 = 0.00633015 - N167*0.285904 - N167*N26*1.04882 + N167^2*0.54392 + N26*1.28199 + N26^2*0.500316

N26 = -0.00329322 + N36*0.533516 + N47*0.46977

N47 = -0.146834 + N475*0.184454 - N475*N64*0.235955 + N64*1.1378 + N64^2*0.0740878

N64 = 0.00165014 - N86*N125*0.351028 + N86^2*0.352612 + N125*0.995084

N125 = -0.11794 + N289*1.06271 + N289*N212*0.406202 - N289^2*0.496514 + N212*0.174208

N212 = 0.109145 + N305*N337*0.254905 + N337*0.775876 - N337^2*0.157401

N289 = 0.357361 - N439*0.320034 + N439*N532*1.34373 - N532*0.493073 + N532^2*0.102772

N86 = -0.0853159 + N341*0.51966 - N341*N145*0.460498 + N341^2*0.117391 + N145*0.620965 + N145^2*0.287804

N145 = 0.134634 + N309*0.72817 + N309*N332*0.600665 - N309^2*0.347556 - N332^2*0.125878

N332 = 0.354385 - N463*0.512867 + N463*N495*0.895426 + N463^2*0.219164

N495 = 0.24536 + N567*N585*1.07204 - N585*0.616472 + N585^2*0.287297

N567 = 0.727256 + "Insolation(B), arctan"*1.46604 - "Insolation(B), arctan"*""Wind Speed(NB)", arctan"*0.45406 - "Insolation(B), arctan"^2*0.609979 - ""Wind Speed(NB)", arctan"^2*0.475217

N463 = 1.18243 - N521*2.122 + N521*N570*1.40412 + N521^2*0.834894 - N570*0.370155

N341 = -0.0281012 + N466*N503*0.928183 + N503*0.20344 - N503^2*0.103289

N503 = 0.334785 + N566*N580*1.06962 - N580*0.820481 + N580^2*0.399752

N580 = 1.50914 - ""Distance from Grid(NB)", arctan"*1.01155 + ""Distance from Grid(NB)", arctan"*"Humidity(NB), arctan"*0.779795 - "Humidity(NB), arctan"^2*0.97292

N466 = 1.40947 - N549*1.22434 + N549*N552*2.12501 - N552*1.32646

N549 = 1.3105 + ""Sunlight Hours(B)", arctan"*0.350157 - ""Cost of Labours(NB)", arctan"*1.07131

N475 = 0.0102089 + N521*N577*0.985985

N521 = 1.23103 + ""No. of Consumers(B)", arctan"*0.628912 - ""Cost of Labours(NB)", arctan"*1.50238 + ""Cost of Labours(NB)", arctan"^2*0.562896

N36 = -0.0437058 + ""Sunlight Hours(B)", arctan"*0.0546366 + ""Sunlight Hours(B)", arctan"^2*0.100037 + N51*0.996476

N51 = -0.0412742 + N111*1.05374 - N111*N133*1.60318 + N111^2*0.596852 + N133^2*0.989285

N133 = -0.0761002 - N217*N267*1.71468 + N217^2*1.08019 + N267*1.1167 + N267^2*0.590211

N267 = -0.0573265 + N510*0.368356 + N282*0.537232 + N282^2*0.119767

N282 = 0.703714 - N437*0.96169 + N437*N517*1.27682 + N437^2*0.250698 - N517*0.316712

N517 = -1.69121 + N578*0.946539 + N584*2.4697 - N584^2*0.698387

N437 = -0.051941 + N509*N577*1.05021

N510 = -0.577487 + N566*1.04029 + N591^2*0.518822

N566 = 0.417699 + "Insolation(B), arctan"*0.9935 - "Insolation(B), arctan"*""Rate of Energy(B)", arctan"*0.41444 + ""Rate of Energy(B)", arctan"^2*0.926692

N217 = 0.10231 + N270*N334*0.333686 + N334*0.817616 - N334^2*0.264655

N334 = 0.736708 - N448*0.50996 + N448*N526*1.48084 - N526*0.986595 + N526^2*0.252213

N526 = -1.27002 + N588*2.25766 + N588*N591*0.430274 - N588^2*0.763451 + N591^2*0.360636

N448 = 0.00534718 + N511*N574*0.991538

N574 = 1.66683 - "Humidity(NB), arctan"*0.165426 + "Humidity(NB), arctan"*""Installation Cost(NB)", arctan"*1.12719 - "Humidity(NB), arctan"^2*0.921521 - ""Installation Cost(NB)", arctan"*1.57678 + ""Installation Cost(NB)", arctan"^2*0.394245

N270 = 0.529048 - N439*0.706275 + N439*N516*1.21261 + N439^2*0.165753 - N516*0.234712

N516 = -0.636895 + N578*1.2712 - N578^2*0.133451 + N588^2*0.48593

N111 = -0.00102201 + N166*0.99102 - N166*N264*0.252785 + N264^2*0.258827

N264 = 0.126321 + ""Rate of Energy(B)", arctan"*N354*0.582548 + N354*0.515453 + N354^2*0.0757286

N354 = 0.107713 - N471*0.0239214 + N471*N498*0.905296

N471 = 1.50681 - N522*1.59952 + N522*N575*2.15015 + N522^2*0.175109 - N575*1.25732

N166 = 0.161915 + N299*0.319857 + N299*N309*0.141733 + N309*0.343439

N167 = 0.198966 + N309*0.59638 + N309*N346*0.489322 - N309^2*0.231749 - N346^2*0.0810712

N346 = 0.294561 - N443*0.480679 + N443*N527*1.43786 - N527^2*0.244257

N527 = -0.0135168 + N584*N591*1.01191

N591 = 1.19533 - ""Distance from Grid(NB)", arctan"*0.957873 + ""Distance from Grid(NB)", arctan"^2*0.319908 + ""Sunlight Hours(B)", arctan"*0.302055

N443 = -0.0377317 + N511*N577*1.03812

N577 = 1.25142 - ""Installation Cost(NB)", arctan"*1.29881 + ""Installation Cost(NB)", arctan"^2*0.633812 + ""Rate of Energy(B)", arctan"^2*0.634072

N511 = 1.12323 + "Insolation(B), arctan"*0.762002 - ""Cost of Labours(NB)", arctan"*1.04531

N309 = 0.22448 - N438*0.537838 + N438*N532*1.13164 + N438^2*0.154814

N438 = 0.508976 - N522*0.83837 + N522*N551*1.71175 - N551^2*0.368959

N534 = -1.86665 + N584*1.93069 - N584*N589*0.945033 + N589*1.88092

N584 = 0.797686 + ""No. of Consumers(B)", arctan"*0.759187 - ""No. of Consumers(B)", arctan"*""Wind Speed(NB)", arctan"*1.28066 + ""No. of Consumers(B)", arctan"^2*0.478609

N439 = 0.863901 - N523*1.61219 + N523*N551*2.09943 + N523^2*0.196996 - N551^2*0.539163

N551 = 1.9288 - ""Distance from Grid(NB)", arctan"*1.38504 + ""Distance from Grid(NB)", arctan"*""Installation Cost(NB)", arctan"*1.64173 - ""Installation Cost(NB)", arctan"*1.48715

N523 = 1.73038 + ""Wind Speed(NB)", arctan"*""Cost of Labours(NB)", arctan"*1.11678 - ""Wind Speed(NB)", arctan"^2*1.16927 - ""Cost of Labours(NB)", arctan"*1.51521

N416 = 0.0161243 + N493*N545*1.00486 - N545*0.0264288

N545 = 0.011443 + N587*N589*0.986268

N587 = 1.52414 + ""Wind Speed(NB)", arctan"*"Humidity(NB), arctan"*0.925641 - ""Wind Speed(NB)", arctan"^2*1.1854 - "Humidity(NB), arctan"*0.927134

N493 = -0.128177 - ""Cost of Labours(NB)", arctan"*N570*1.48309 + ""Cost of Labours(NB)", arctan"^2*0.543497 + N570*1.63232

N77 = -0.172545 + N363*0.245239 - N363*N116*0.684454 + N363^2*0.187907 + N116*1.11069 + N116^2*0.341058

N116 = -0.0201301 + N208*0.57187 + N222*0.448218

N222 = 0.0728594 + N305*0.387646 + N305*N313*0.0813176 + N313*0.439898

N313 = 0.467077 - N468*0.77705 + N468*N498*1.33969 + N468^2*0.155429 - N498^2*0.199502

N498 = -0.0929565 + N565*N585*1.09464

N585 = 1.30711 - ""Sunlight Hours(B)", arctan"*""Installation Cost(NB)", arctan"*0.299154 + ""Sunlight Hours(B)", arctan"^2*0.500446 - ""Installation Cost(NB)", arctan"*1.17538 + ""Installation Cost(NB)", arctan"^2*0.59571

N565 = 0.292736 + "Insolation(B), arctan"*1.02844 - "Insolation(B), arctan"*""No. of Consumers(B)", arctan"*0.609497 + ""No. of Consumers(B)", arctan"*0.865128

N468 = 0.7359 - N522*1.13 + N522*N570*1.2864 + N522^2*0.374011 - N570*0.30172

N570 = 1.1819 - ""Distance from Grid(NB)", arctan"*1.28587 + ""Distance from Grid(NB)", arctan"^2*0.647262 + ""Rate of Energy(B)", arctan"^2*0.799963

N522 = 1.87363 - "Humidity(NB), arctan"*0.490671 + "Humidity(NB), arctan"*""Cost of Labours(NB)", arctan"*1.83514 - "Humidity(NB), arctan"^2*0.919382 - ""Cost of Labours(NB)", arctan"*1.832

N305 = 0.705982 - N431*1.09585 + N431*N539*1.16464 + N431^2*0.341687 - N539*0.202861 + N539^2*0.0244017

N431 = 0.061061 - N502*0.0569678 + N502*N575*0.988732

N208 = 0.113504 + N299*0.777634 + N299*N337*0.339689 - N299^2*0.246152

N337 = 0.778645 - N430*1.103 + N430*N539*1.22235 + N430^2*0.339541 - N539*0.451309 + N539^2*0.143392

N539 = -1.11186 + N578*1.03575 + N597*1.07376

N597 = 0.680317 + ""Sunlight Hours(B)", arctan"*0.427727 - ""Sunlight Hours(B)", arctan"*""Rate of Energy(B)", arctan"*0.1423 - ""Sunlight Hours(B)", arctan"^2*0.115228 + ""Rate of Energy(B)", arctan"^2*0.750536

N578 = 0.884052 + "Insolation(B), arctan"*0.785828 - "Humidity(NB), arctan"*0.51441

N430 = 0.0859863 + N509*N560*1.05427 - N560*0.145036

N299 = -0.0153284 + N436*N532*1.01331

N532 = -1.78622 + N588*1.74634 - N588*N589*0.719186 + N589*1.75601

N588 = 0.641017 + ""No. of Consumers(B)", arctan"*0.108875 + ""No. of Consumers(B)", arctan"^2*0.580412 + ""Rate of Energy(B)", arctan"^2*0.657429

N436 = 0.0737964 + N509*N552*1.85505 - N509^2*0.4175 - N552^2*0.459083

N552 = 2.05943 - ""Wind Speed(NB)", arctan"*1.42788 + ""Wind Speed(NB)", arctan"*""Installation Cost(NB)", arctan"*1.87248 - ""Installation Cost(NB)", arctan"*2.06915 + ""Installation Cost(NB)", arctan"^2*0.500087

N509 = 1.94205 - ""Distance from Grid(NB)", arctan"*1.21151 + ""Distance from Grid(NB)", arctan"*""Cost of Labours(NB)", arctan"*1.31302 - ""Cost of Labours(NB)", arctan"*1.57474

N363 = -0.00289997 - N442*0.0393829 + N442*N547*1.03961

N547 = -0.0293643 + N589*N593*1.02873

N593 = 0.969854 - "Humidity(NB), arctan"^2*0.621906 + ""Rate of Energy(B)", arctan"^2*0.692014

N589 = 0.56979 + "Insolation(B), arctan"*0.759103 + ""Sunlight Hours(B)", arctan"*0.247712

N442 = 0.0146554 + N502*N569*0.97991

N569 = 1.11251 + ""No. of Consumers(B)", arctan"*0.22887 + ""No. of Consumers(B)", arctan"^2*0.500151 - ""Distance from Grid(NB)", arctan"*1.0271 + ""Distance from Grid(NB)", arctan"^2*0.38936

N502 = 2.18949 - ""Cost of Labours(NB)", arctan"*2.01486 + ""Cost of Labours(NB)", arctan"*""Installation Cost(NB)", arctan"*2.31912 - ""Installation Cost(NB)", arctan"*1.74436

N364 = 0.376231 + N454*N485*0.849927 - N485*0.25838

N485 = 2.15669 - N560*3.01202 + N560*N575*2.53623 + N560^2*0.644717 - N575*1.16147 - N575^2*0.199675

N575 = 1.73578 - ""Distance from Grid(NB)", arctan"*1.14475 + ""Distance from Grid(NB)", arctan"*""Wind Speed(NB)", arctan"*1.08827 - ""Wind Speed(NB)", arctan"*1.03588

N560 = 0.917157 + ""No. of Consumers(B)", arctan"*0.469994 - ""No. of Consumers(B)", arctan"*""Installation Cost(NB)", arctan"*1.45061 + ""No. of Consumers(B)", arctan"^2*0.991567 - ""Installation Cost(NB)", arctan"*0.193698

N454 = 0.434562 - N518*1.00827 + N518*N556*1.04593 + N518^2*0.487916

N556 = 0.881592 + "Insolation(B), arctan"*0.944798 - "Insolation(B), arctan"*""Distance from Grid(NB)", arctan"*0.420568 - ""Distance from Grid(NB)", arctan"*0.510303

N518 = 1.27506 - ""Cost of Labours(NB)", arctan"*1.06458 + ""Rate of Energy(B)", arctan"^2*0.714219

References

Khabibrakhmanov I, Lu S, Hamann HF, Warren K (2016) On the usefulness of solar energy forecasting in the presence of asymmetric costs of errors. IBM J Res Dev 60(1):1–7

Mirzaei S, Avakhdarestani S (2016) Development of failure mode and effects analysis using fuzzy analytical network process. Int J Prod Qual Manag 17(2):215–235

Munsell M (2015) The US installed 6.2GW of solar in 2014, up 30 % over 2013, Retrieved from http://www.greentechmedia.com/articles/read/the-us-installed-6.2-gw-of-solar-in-2014-up-30-over-2013 on 11 Feb 2016

Solar Energy Industries Association (2015) Solar industry data, Retrieved from http://www.seia.org/research-resources/solar-industry-data on 11 Feb 2016

Stökler S, Schillings C, Kraas B (2016) Solar resource assessment study for Pakistan. Renew Sustain Energy Rev 58:1184–1188

Chapter 7
Conclusion

Abstract In the conclusion of the study it was found that the methodology depicted in the present investigation for site selection for solar power plants is both objective, cognitive and relative in nature. The indicator is a stand alone model which can be mebedded anywhere to predct site suitability real time. But dependence on type of method used is a major drawback which need to be addressed in the later research works.

Keywords Real time system · On field analysis · Cognitive

The present investigation was an attempt to device a system to represent the potentiality of a location for installation of solar power plants. In this regard the AHP and ANP MCDM method was used to find the importance of the input variables. The GMDH neural network was utilized to find the interrelation between the input variables and the output indicator.

The indicator was developed in such a manner that it becomes directly proportional to the beneficiary and indirectly proportional to the non-beneficiary parameters with respect to the study objective. The indicator was the ratio of weighted sum of all the beneficiary parameters and non-benficiary parameters respectively placed in the numerator and denominator.

The indicator was utilized to find the feasibility of 12 different locations collected from all the five continents of the World.

According to the results it was found that Insolation of a region is the most important parameter for identification of sites for installation of solar power plant. The weather parameter humidity was found to be minimum important.

Among the two MCDM method used, the model developed with ANP method was found to be most optimal if the output data of the model is transformed by Arc Tangent function and the neural function is non-linear. Model will be required to be trained with the GMDH algorithm.

The model developed in the manner described in the earlier paragraph was found to have the maximum EPM among all the 24 developed models for the present investigation.

© The Author(s) 2016   47
M. Majumder and A.K. Saha, *Feasibility Model of Solar Energy
Plants by ANN and MCDM Techniques*, SpringerBriefs in Energy,
DOI 10.1007/978-981-287-308-8_7

When the indicator was used to find the feasibility of the 12 different location selected from all over the World it was found that New York has the highest whereas Paris has the lowest suitability for installation of solar power plant.

Among the countries which have added capacity to its existing solar energy production USA is ranked third below China and Japan which depicts the reliability of the result derived from the indicator which is referred as S-Value (IEA 2014). No city from China and Japan was considered in the selected 12 cities. That is why city of New York becomes the most feasible city for installation of solar power plant.

7.1 Strength

The indicator developed for the present study is derived by an objective method. The model developed for prediction of the indicator provides a cognitive ability to the same. As the importance of the variables and also the parameters were selected based on a literature survey and by an objective method the problem of biasness does not arise in this case.

The alternative as well as criteria selected for the study was a result of the metastatic analysis conducted within various literatures related to study objective.

As the neural model yield a standalone equation to predict the indicator, this equation can be embedded in portable devices to conduct in situ surveys. The model can also be used to perform scenario analysis of different conditions that may arise in the future.

The future prediction of location feasibility will also ensure profitability and maximum utilization of resources even in the future years.

7.2 Limitation

The dependence of the model on the methods and parameters used to develop is one of the major limitations of the present model. New parameters or new method can change the model results.

Some specific mechanism must be derived so that while comparing locations the type of method and parameters must not be changed.

7.3 Future Scope

The developed indicator can be modified to include the impact of urbanization and climate change both of which are eminent in the near future.

The model input parameters were compared with respect to technical, economical and locational criteria. But the impact of environmental and social criteria may also be incorporated for a more extensive priority analysis.

Reference

International Energy Agency (IEA) (2014) Snapshot of global PV market, Retrieved from http://www.iea-pvps.org/fileadmin/dam/public/report/technical/PVPS_report_-_A_Snapshot_of_Global_PV_-_1992-2014.pdf on 11 Feb 2016